Chemical Reactions and Processes under Flow Conditions

RSC Green Chemistry

Series Editors:
James H. Clark, *Department of Chemistry, University of York, York, UK*
George A. Kraus, *Department of Chemistry, Iowa State University, Iowa, USA*

Titles in the Series:
1: The Future of Glycerol: New Uses of a Versatile Raw Material
2: Alternative Solvents for Green Chemistry
3: Eco-Friendly Synthesis of Fine Chemicals
4: Sustainable Solutions for Modern Economies
5: Chemical Reactions and Processes under Flow Conditions

How to obtain future titles on publication:
A standing order plan is available for this series. A standing order will bring delivery of
each new volume immediately on publication.

For further information please contact:
Book Sales Department, Royal Society of Chemistry,
Thomas Graham House, Science Park, Milton Road, Cambridge,
CB4 0WF, UK
Telephone: + 44 (0)1223 420066, Fax: + 44 (0)1223 420247, Email: books@rsc.org
Visit our website at http://www.rsc.org/Shop/Books/

Chemical Reactions and Processes under Flow Conditions

Edited by

S.V. Luis and E. Garcia-Verdugo
Associated Unit for Advanced Organic Materials, University Jaume I/CSIC, Castellón, Spain

RSCPublishing

RSC Green Chemistry No. 5

ISBN: 978-0-85404-192-3
ISSN: 1757-7039

A catalogue record for this book is available from the British Library

Published by The Royal Society of Chemistry,
Thomas Graham House, Science Park, Milton Road,
Cambridge CB4 0WF, UK

Registered Charity Number 207890

For further information see our web site at www.rsc.org

Preface

Many industrial processes involve continuous catalytic reactors for manufacturing chemicals, petrochemicals, fuels, *etc*. New reactor designs and modifications of existing ones allow improvements to yields and time operation, while reducing costs. The purpose of this book is to provide an overview of chemical reactions and processes that make use of continuous reactors, utilising a variety of catalysts ranging from organic to inorganic materials, and with emphasis on some of the engineering factors governing them. To do this, experts in different areas present selected highlights deemed of particular importance.

In the first chapter, key engineering concepts are described with detail, together with the main parameters and approaches for the design of efficient continuous reactors. This introductory chapter is followed by two chapters devoted to the use of functional organic and inorganic catalysts working under flow conditions. Functional organic materials constitute a combination that ranges from carbon nanotubes to self-assembled polymers, encompassing catalytic applications that will include the preparation of chiral materials, *etc*. In a similar way, the importance of zeolites and molecular sieves as catalysts and their implementation in petrochemical and oil refining process are treated in a separate chapter. Here, classic examples of continuous processes such as catalytic cracking and the isomerisation of alkanes and alkenes, among others, are described in detail. In a separate section, the use of zeolites and molecular sieves is extended to the manufacture of organic intermediates and fine chemicals–an area of interest, owing to environmental legislation, that has stimulated the development of cleaner methodologies, *i.e.* heterogeneous catalysts within flow processes. Continuous catalytic processes in which the catalyst is present within an ionic liquid or in supercritical fluids are also considered, and Chapter 5 deals with recent advances in catalytic flow processes where an ionic liquid and/or supercritical fluid have replaced organic solvents.

RSC Green Chemistry No. 5
Chemical Reactions and Processes under Flow Conditions
Edited by S.V. Luis and E. Garcia-Verdugo
© The Royal Society of Chemistry 2010
Published by the Royal Society of Chemistry, www.rsc.org

In the past few years, advances in microreactors and microfluid devices have demonstrated that the miniaturisation of chemistry can introduce significant advantages with respect to cost, safety, throughput, kinetics and process scale-up. These research efforts devoted to execute a synthesis in a microreactor are not new, but started in the 1970s. Nonetheless, in recent years, there has been a renewed interest in microreactors and associated operations in response to pressures from two separate but related disciplines—preclinical drug discovery, and chemical development and manufacturing. Thus microreactors and microfluid systems are described in Chapter 4, where the reader can find information on important aspects of the design of microreactors, the main advantages associated with the use and implementation of microfluid devices, and the chemistry performed with these devices.

We hope that this book will be of interest to those chemists engaged in organic synthesis and catalysis in industrial and academic laboratories who are concerned with research and development, as well as graduate students who are interested in this area.

Avelino Corma
Valencia

Contents

RSC Green Chemistry No. 5
Chemical Reactions and Processes under Flow Conditions
Edited by S.V. Luis and E. Garcia-Verdugo
© The Royal Society of Chemistry 2010
Published by the Royal Society of Chemistry, www.rsc.org

Chapter 4   Microfluidic Devices for Organic Processes   118
*Paola Laurino, Arjan Odedra, Xiao Yin Mak Tomas
Gustafsson, Karolin Geyer and Peter H. Seeberger*

Engineering Factors for Efficient Flow Processes in Chemical Industries

ALEXEI A. LAPKIN AND PAWEL K. PLUCINSKI

Centre for Sustainable Chemical Technologies, Department of Chemical Engineering, University of Bath, Bath BA2 7AY, UK

1.1 Introduction

Continuous chemical processes integrated *via* energy and material flows are forming the basis of a highly successful petrochemical industry. Effectively all petrochemical processes, starting from crude oil heating, hydrotreating, cracking, refining and further synthesis of bulk products are performed in continuous flow reactors and separators. The same applies to other large-scale processes, for example, the synthesis of ammonia and sulfuric acid. The scale of production and the close integration of materials and energy are the key attributes of traditional continuous flow processes that contribute to their remarkable efficiency.

The introduction of continuous flow processes in smaller-scale manufacturing such as speciality chemicals, chemical intermediates, pharmaceutical intermediates, active ingredients in agrochemicals and pharmaceuticals, nutraceuticals, fragrances, surfactants, *etc.* faces significant challenges due to the reliance of these industries on sunken capital—the existing infrastructure of batch multipurpose plants and the slow introduction of the suitable scale technologies. Only recently have the compact and microreactor systems been developed that

RSC Green Chemistry No. 5
Chemical Reactions and Processes under Flow Conditions
Edited by S.V. Luis and E. Garcia-Verdugo

Published by the Royal Society of Chemistry, www.rsc.org

could begin to replace the traditional batch multipurpose plants.[1–16] However, the advantages of continuous processing are clear enough. The processes are generally more efficient than batch ones and offer much higher throughput per unit volume and per unit time. Reactants are introduced continuously, react on contact within a smaller reaction space with better defined temperature and flow fields, and are removed continuously from the reaction space. There is better control of process variables and the risk of side reactions is reduced. The reactor volume is determined by the flow rate and residence time of the materials rather than vice versa; therefore, vessels can be smaller and heat transfer and mixing are easier to control. Waste levels are generally also lower.

The areas in which flow processes have been developing at a rapid pace are biotechnology and biomedicine.[17,18] In these areas, the closer relation to living systems (which can be said to be 'flow systems' and highly material/energy integrated systems) gives stronger impetus to the exploitation of the functionality of the flow reactors.[19] Such features of small continuous flow systems as the extensive use of *in situ* analytics,[20] sequential operations, use of weaker fields such as electric and magnetic separations,[21] microwave heating[22] and sonication as well as parallelisation and automation for increase in productivity have already found numerous applications in biotechnology and are rapidly penetrating into chemical processes.

This chapter considers the engineering basis for the design of continuous flow chemical and biochemical reactors at different scales. The emphasis is on new and emerging areas of process intensification (PI), flow chemistry, and compact and microreactors; process engineering of petrochemical reactors is well covered in earlier literature and some aspects are discussed in Chapter 3.[23,24] One of the main differences between large-scale and micro-scale flow processes, to which we pay particular attention, is the more significant role of surface–fluid interactions and hence the need to account for solid–fluid physico-chemical interactions in the reactor design. The issues of scale up of small-scale flow reactors are also considered.

The process intensification concept that emerged in industry initially aimed to reduce the physical footprint of plants, and hence reduce capital investment and improve safety.[15] This concept is now widely accepted in the broader meaning of the reduction in the overall impact of chemical processes over their entire life cycle. The different tools of PI are shown in Figure 1.1.

In flow chemistry, a chemical reaction is run in a continuously flowing stream; liquids (normally reagent/substrate solutions) are driven through a reactor which is often a capillary or tubing. In recent years, flow chemistry has emerged as a viable means for performing many types of chemical transformations. Within industry, flow chemistry is already having a major impact: large pharmaceutical companies have teams of chemists and chemical engineers active in the field. On the macro scale, flow processes are being developed for the manufacture of active pharmaceutical ingredients where a series of synthesis reactions, work-up steps and crystallisation of the final active pharmaceutical intermediate (API) are performed in a sequence of flow modules as shown schematically in Figure 1.2.

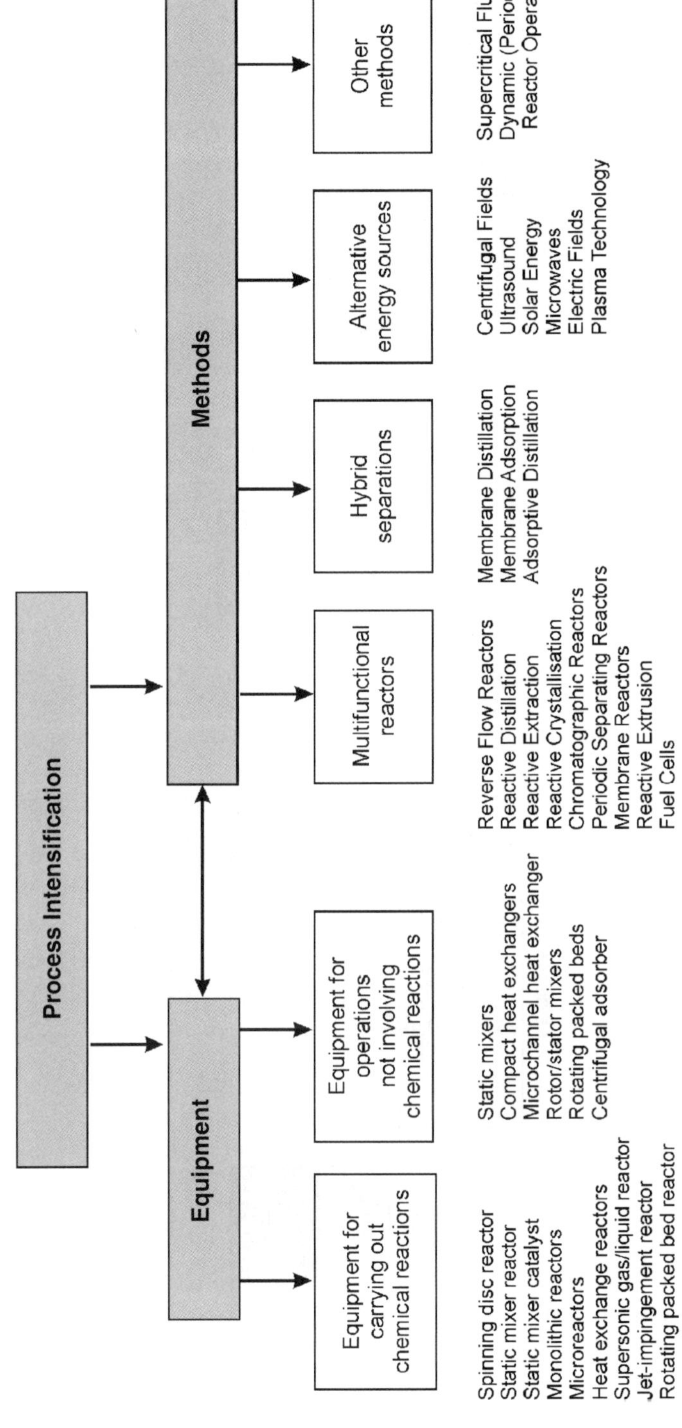

Figure 1.1 The concept of process intensification (PI) and different tools of PI (adopted from ref. 15).

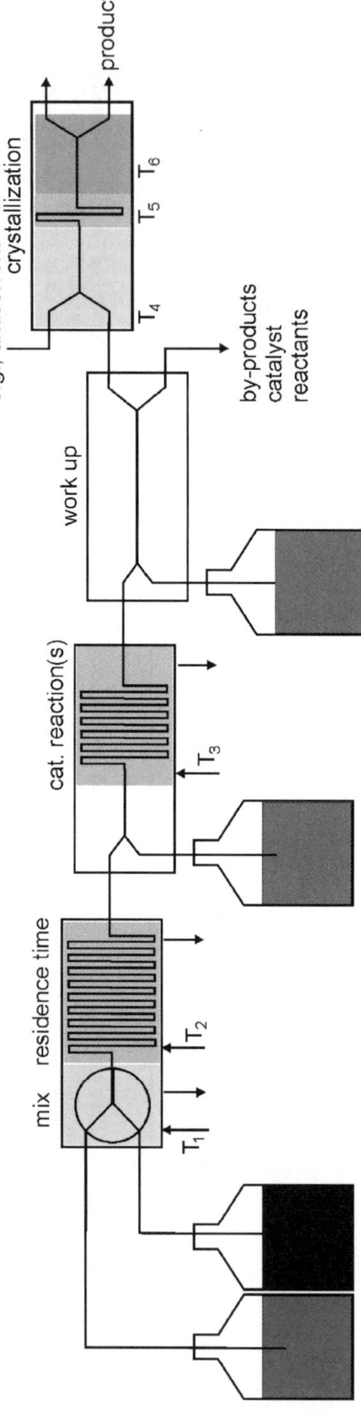

Figure 1.2   Schematic representation of a modular flow chemistry kit based on multifunctional flexible units.

1.2 Heterogeneous Catalytic Flow Processes in the Petrochemical Industry: A Brief Overview

1.2.1 Gas–solid and Liquid–solid Catalytic and Non-catalytic Continuous Processes

Reactions in systems where at least one reactant is solid play a major role in the materials processing industries, encircling a broad range of operations such as extractive metallurgy (*e.g.* ore leaching), coal gasification (or more generally combustion of solid fuels: coal, lignite, *etc.*), pyrolysis of lignocellulosic products, incineration of municipal waste and catalyst regeneration.[25,26] Most of these reactions can be represented by a general stoichiometric equation:

$$a\ A_{(fluid)} + b\ B_{(solid)} \rightarrow c\ C_{(solid)} + d\ D_{(fluid)}$$

The reactions involving a solid reactant include the following elementary steps (Figure 1.3, shown here as an example of a gas–solid system with solid particle pyrolysis):

  (i) external (gas phase) mass transfer;
 (ii) diffusion inside the pores (if solid is porous);
(iii) chemical reaction(s) between gaseous and solid reactants (may involve adsorption of reactant(s) and desorption of reaction products);
(iv) diffusion or reaction(s) product(s) from the reaction site towards the external surface of the solid;
 (v) external mass transfer of formed reaction product(s) away from the solid interface.

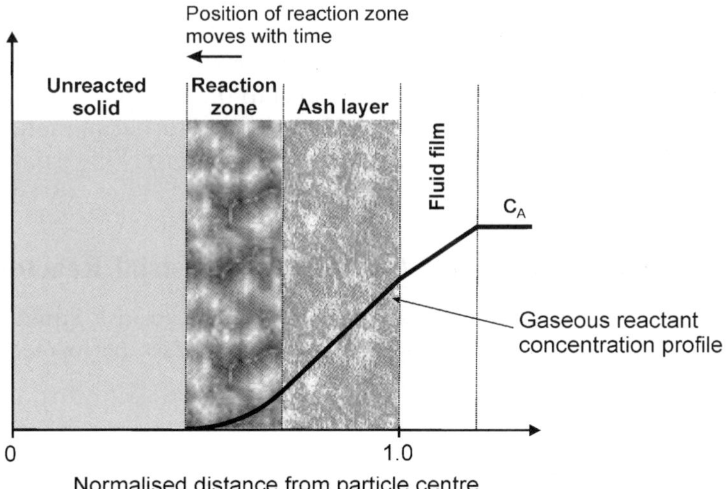

Figure 1.3 Basic steps of solid–fluid reactions (adopted from ref. 26).

The diffusion of reaction products through the pore system of a solid material and external mass transfer—forming an integral part of the process—are important if the reaction is reversible. Although the process steps listed above occur in series, any one or more of these could be rate limiting.

In slow reacting systems, the overall dynamics will be limited by the surface kinetics (intrinsic rate); the increase of reaction rate may change the limitation to the pore diffusion. For faster exothermic reactions, the temperature gradient across the particle or fluid film might become the controlling factor. In the case of very fast chemical reactions, the mass transfer in the external fluid film becomes the rate limiting step.

An important difference that distinguishes fluid–solid reactions from their catalytic counterparts is that, in non-catalytic systems, a solid is also involved as a reactant. Continuous consumption of the solid phase during the reaction leads to structural changes of the solid bed morphology, and the reactor system is always in the transient state.

The rate of the overall process for external mass transfer limitation can easily be obtained from the knowledge of mass transfer around solid particles and several correlations for fixed or moving solid particles are reported in the literature.[25,26] The challenges in the mathematical description of these types of reactors concern the molecular diffusion in the pore systems of a solid phase. Continuous changes of a solid's morphology (pore shrinkage or closure, swelling, sintering, softening or cracking of the particles) affect the effective diffusivity with the progress of reaction.

The primary consideration in the design and analysis of such systems is the mode of contact of the phases. Fixed, fluidised and moving bed techniques appear to be the most common mode of phase contacting. Horizontal moving bed, pneumatic conveyers, rotating cylinders and flat hearth furnaces are less common.

In catalytic fluidised bed reactors, the problems of inhomogeneity of the fluidised bed when the gas phase is used as a fluidising agent could be overcome by using an external magnetic force and magnetisable catalyst particles.[27,28] Fluidisation of magnetisable particles by a gas stream in the presence of a uniform applied magnetic field oriented parallel to the flow prevents the hydrodynamic instability that otherwise leads to bubbles and turbulent motion within the medium. The fluidised emulsion phase expands uniformly in response to gas flow velocity.

1.2.2 Two-phase Gas–liquid Continuous Industrial Reactors

Chemical reactions between a gas and a solute dissolved in a liquid are very common in industry.[29] Examples of important processes performed in gas–liquid reactors include:

- absorption of acid gases;
- oxidation of organic compounds by oxygen or air;
- chlorination;
- hydrogenation of organic compounds.

In such reactions, a gaseous component(s) is dissolved in the liquid phase where it reacts with other reagent(s). In the catalytic reactions (homogeneous catalysis), the liquid phase contains a catalyst together with liquid reactant(s). Slightly different scenarios may occur for a biphasic (liquid–liquid) mode of operation. For example, a liquid reagent will be dissolved in the other liquid phase containing the catalyst.

The fundamental analysis of two-phase reactors is complex due to the coupling of simultaneously occurring diffusion and reactive processes. In addition, the hydrodynamic conditions of the reactive two-phase system are difficult to define.[29]

For a chemical reaction taking place in the laminar film and bulk liquid, starting from elementary mass balance of a reactant A, the expression for calculating the overall reaction rate can be developed as shown in eqn (1.1):

$$N_A(z=1) = \frac{k_{Al}Ha}{tanh(Ha)} \left\{ \frac{\frac{k_{Ag}\,tanh(Ha)}{k_{Al}\cdot Ha}p_A + \frac{c_{Ab}}{cosh(Ha)}}{cosh(Ha)\left[1 + \frac{k_{Ag}H\,tanh(Ha)}{k_{Al}\cdot Ha}\right]} - c_{Ab} \right\} \tag{1.1}$$

where: c_{Ab} = bulk concentration of A in the liquid phase; H = Henry constant; Ha = Hatta number; k_{Ag} = mass transfer coefficient in the gas phase; k_{Al} = mass transfer coefficient in the liquid phase; p_A = partial pressure of A in the gas phase; x = distance from the interface; z = dimensionless length $(z = \frac{x}{\delta_l})$; δ_l = thickness of the laminar layer.

Hatta number, or more precisely Ha^2, is a dimensionless number being a ratio of the maximum rate of the reaction in the liquid laminar film and the maximum rate of transport through the liquid film. For a first order chemical reaction is defined in eqn (1.3) as follows:

$$Ha = \delta_l\sqrt{\frac{k_A}{D_{Al}}} = \sqrt{\frac{D_{Al}k_A}{k_{Al}}} \tag{1.2}$$

where: D_{Al} = diffusion coefficient of A in the liquid phase; k_A = reaction rate constant; k_{Al} = mass transfer coefficient in the liquid phase.

For analysis of such coupled fluid–fluid systems (which may include two liquid phases), it is useful to distinguish between three regimes of reaction rate which are characterised by different Ha values and the enhancement factor *E* (Table 1.1). The mass transfer rate between two phases is compared with that for pure physical adsorption *via* enhancement factor (*E*) as shown in eqn (1.3):

$$E = \frac{\text{rate of reaction or flux of A}}{\text{maximum rate of mass transfar of A through liquid film}} \tag{1.3}$$

For slow reactions (Ha < 0.3), the overall rate of mass transfer is not enhanced by the chemical reaction (which takes place mainly in the bulk of reaction phase), and the enhancement factor becomes approximately 1.[30] For the

Table 1.1   Regimes of mass transfer/reaction limitations for fluid–fluid reactive systems.

For Ha < 0.3, $E \approx 1$, $f \approx 1$	**Regime 1**: slow reactions, controlled by chemical kinetics.
	Rate of chemical reaction $<$ rate of mass transfer
For Ha < 0.3, $E \approx 1$, $f \approx 0$	**Regime 1**: slow reactions, controlled by diffusion.
	Rate of chemical reaction $>$ rate of mass transfer
For $0.3 <$ Ha < 3, $E > 1$, $f \approx 0$	**Regime 2**: fast reactions
	Rate of chemical reaction $>$ rate of mass transfer
For Ha > 3, $E =$ Ha, $f \approx 0$	**Regime 3**: Instantaneous reactions
	Rate of chemical reaction $\gg$ rate of mass transfer

where: $f = \frac{c_A}{H p_A}$ (c_A = concentration of A; H = Henry's constant)

Table 1.2   Orders of magnitude of mass transfer parameters of various two-phase reactors (adopted from ref. 25).

	Liquid hold-up [%]	Gas hold-up[a] [%]	k_G [$m\,s^{-1}$]	k_L [$m\,s^{-1}$]	Interfacial area[a] [$m^2\,m^{-3}$]
Bubble column	>70	2–30	$(1\text{–}5) \times 10^{-2}$	$(1\text{–}5) \times 10^{-4}$	100–500
Mechanically stirred units	>70	2–30	$(1\text{–}5) \times 10^{-2}$	$(1\text{–}6) \times 10^{-4}$	200–2000
Plate columns	60–80 10–20[b]	20–40 80–90[b]	$(1\text{–}5) \times 10^{-2}$	$(1\text{–}5) \times 10^{-4}$	200–500 25–100[b]
Packed columns	5–15	50–80	$(1\text{–}5) \times 10^{-2}$	5×10^{-5} to 3×10^{-4}	50–250
Scrubbers	$<5^c$ $<30^d$	$>95^c$ $>70^d$	$(1\text{–}5) \times 10^{-2}$	$(1\text{–}5) \times 10^{-4}$	25–200

[a]Related to the sum of active volumes of the two phases (gas + liquid).
[b]Related to the total volumes of the column (gas + liquid + volume occupied by internal structures + inactive volumes).
[c]Venturi scrubber.
[d]Spray scrubber.

intermediate range of Hatta number ($0.3 <$ Ha < 3), the overall rate of mass transfer is improved by the chemical reaction ($E = \frac{\text{Ha}}{\tanh(\text{Ha})}$). In the case of high Hatta number (Ha > 3), the reaction is very fast and proceeds only within the laminar boundary layer ($E =$ Ha)

In gas–liquid reactions, yield and selectivity could also be affected by mass transfer, the nature of gas–liquid contact and the residence time distribution. Table 1.2 gives orders of magnitude of mass transfer parameters for various two-phase reactors.

Some typical examples of two-phase (gas–liquid or liquid–liquid) reactors (see Figure 1.4) include:

- stirred tanks reactors;
- bubble or spray columns;

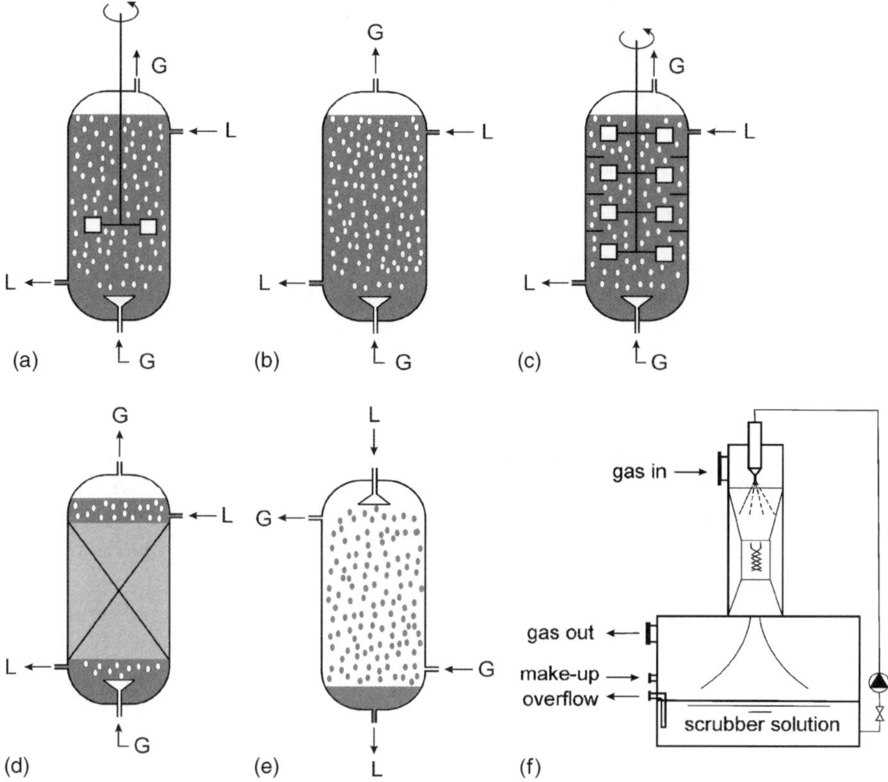

Figure 1.4 Schematic representation of selected industrial gas–liquid reactors (adopted from ref. 29). (a) stirred tank reactor; (b) bubble column; (c) multi-stage bubble column; (d) packed bed column; (e) spray column; (f) Venturi ejector.

- packed columns;
- Venturi reactors

The presence of the homogeneous catalyst mainly influences the rate of chemical reaction; however some other effects may appear if the catalyst has interfacial properties (*e.g.* in micellar and phase transfer catalysis).

1.2.3 Three-phase Catalytic Reactors

Three-phase continuous catalytic processes involving gas, liquid and a solid catalyst are widely used in industrial practice including the manufacturing of commodity chemicals.[25,31,32] The most common example includes liquid phase catalytic hydrogenations, which have been carried out industrially for a very long time.[25] Other process examples[33] include:

- desulfurisation;
- hydrocracking;

- refining of crude oil products in petrochemistry;
- synthesis of butynediol from acetylene and formaldehyde;
- reduction of adiponitrile to hexamethylenediamine.

The liquid phase, often acting as a solvent in such types of reactors, not only dissolves the reactants, but also provides a liquid layer around the catalyst particles, which may help to:

(i) avoid deactivating deposits (*i.e.* guarantee higher catalyst effectiveness);
(ii) achieve better temperature control due to higher heat capacity of liquids; and
(iii) modify active catalytic sites to promote or inhibit certain reaction pathways.[34]

In most applications, the reaction occurs between a dissolved gas and a liquid phase in the presence of a solid catalyst. However, in some cases, when a large heat sink is required for highly exothermic reactions (*e.g.* in the Air Products methanol synthesis process), the liquid is an inert medium and the reaction takes place between the dissolved gases at a solid interface.

In practice, in implementing three-phase reaction systems, several alternatives are available for bringing the three phases into contact. Generally, one can classify these systems based on whether the catalyst is suspended in the reactor (more precisely in the liquid phase—slurry reactors) or is present in the form of a packed bed of catalyst particles (fixed bed reactors).

Reactors with the catalyst dispersed in a liquid phase may exist in three forms: (a) bubble columns, (b) mechanically stirred tanks, and (c) three-phase fluidised beds (see Figure 1.5).

Reactors with the catalyst placed in a fixed bed mode can operate as (a) trickle bed reactors, and (b) packed bubble column reactors. In the first mode,

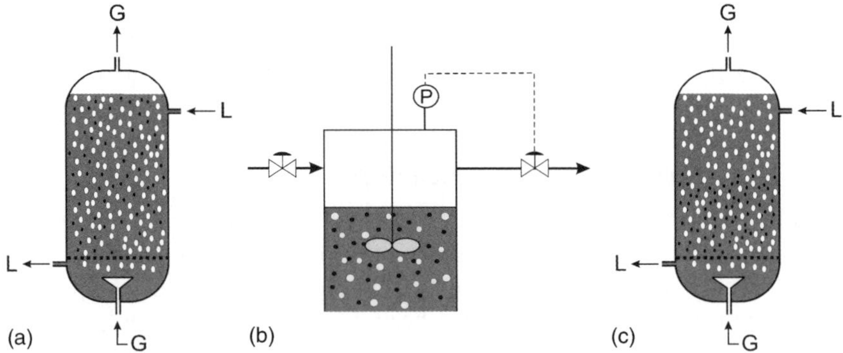

Figure 1.5 Various types of three-phase slurry reactors (adopted from ref. 35). (a) slurry bubble column, counter-current flow; (b) mechanically agitated slurry reactor; (c) three-phase fluidised bed reactor.

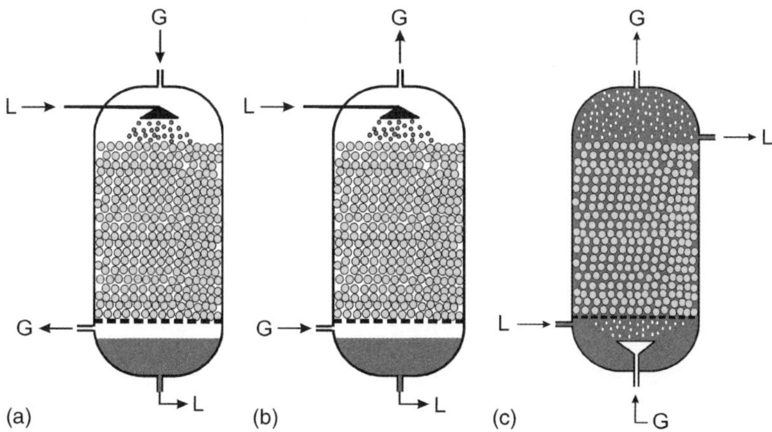

Figure 1.6   Various types of three-phase packed bed reactors (adopted from refs. 35 and 36). (a) trickle bed reactor, co-current flow; (b) trickle bed reactor, counter-current flow; (c) packed-bed reactor, co-current up-flow.

the gas phase comprises the continuous phase of the reactor; in the second, two-phase (gas–liquid) flow occurs through the fixed bed of catalyst particles (see Figure 1.6).

When comparing the various possible reactors offered to users for a given application, it is necessary to consider both the main characteristic features of each type of reactor (Table 1.3) and a number of appreciation criteria of varying importance in order to perform a given reaction in the reactor of choice effectively (Table 1.4).[25]

The modelling and design of three-phase reactors, including the various mass transfer and reaction steps of the process[37] is shown in Figure 1.7.

The first stage of the modelling process considers the following five steps:

(i) Transport from the bulk gas phase to the gas–liquid interface [eqn (1.4)]:

$$-r'_A = k_g a_i \frac{1}{(1-\varepsilon_b)\rho_c} [c_A(g) - c_{Ai}(g)] \ \frac{mol}{g_{cat} \cdot s} \tag{1.4}$$

where: a_i = gas–liquid interfacial area/volume of bed; $c_A(g)$ = bulk gas phase concentration of A; $c_{Ai}(g)$ = interfacial concentration of A; $\frac{mol}{g_{cat} \cdot s}$ = unit of reaction rate r'_A in moles per unit of the mass of catalyst and per time unit; ε_b = bed porosity (gas + liquid); ρ_c = density of catalyst pellet.

(ii) Equilibrium at the gas/liquid interface [eqn (1.5)]:

$$c_{Ai} = \frac{c_{Ai}(g)}{H} \tag{1.5}$$

Table 1.3 Main characteristics of various types of three-phase reactors (adopted from ref. 25).

| Features | Catalyst in suspension | | Fixed bed | | | Three-phase fluidised bed |
	Bubble column	Mechanically stirred tank	Down-flow co-current	Up-flow co-current	Counter-current	
$\varepsilon_P{}^a$	0.01	0.01	0.6–0.7	0.6–0.7	0.5^b	0.1–0.5
$\varepsilon_L{}^a$	0.8–0.9	0.8–0.9	0.05–0.025	0.2–0.3	0.05–0.1	0.2–0.8
$\varepsilon_G{}^a$	0.1–0.2	0.1–0.2	0.2–0.35	0.5–0.1	0.2–0.4	0.05–0.2
d_P [mm]	≤ 0.1	≤ 0.1	1–5	1–5	> 5	0.1–0.5
a_S [m^2 m^{-3}]	500	500	1000–2000	1000–2000	500	500–1000
a_{GL} [m^2 m^{-3}]	100–400	100–400	100–1000	100–1000	100–500	100–1000
η (T = constant)	1	1	< 1	< 1	< 1	≤ 1

[a]The values correspond only to the part of reactor occupied by the catalyst and not the entire reactor.
[b]Value corresponding to special shape of particles.

Table 1.4 Comparison of different types of three-phase reactors (adopted from ref. 25).

Appreciation criteria		Catalyst in suspension	Three-phase fluidised bed	Fixed bed
Characteristics associated with the catalyst	Activity	Highly variable, but possible in many cases to avoid the diffusion limitation found in a fixed bed	Highly variable: intra- and extra-particular mass transfers may significantly reduce the activity, especially in fixed bed	
			Back mixing unfavourable	Plug flow favourable
	Selectivity	Selectivity generally un-affected by transfers	As for activity, transfers may decrease selectivity	
			Back mixing often unfavourable	Plug flow often favourable
	Stability	Catalyst replacement between each batch opera-tion helps to overcome pro-blems of rapid poisoning in certain cases	Possibility of continuous catalyst renewal: the catalyst must nevertheless have good attrition resistance	This feature is essential for fixed bed operation: plug flow may sometimes be favourable, due to establish-ment of a poison adsorption front
Technological characteristics	Cost	Consumption usually depends on the impurities contained in the feed and acting as poisons		Necessarily low catalyst consumption
	Heat exchange	Fairly easy to achieve heat exchange	Possibility of heat exchange in the reactor itself	Generally adiabatic operations
	Design difficulties	Catalyst separation sometimes difficult; possible problems in pumps and exchangers due to the risk of deposit and erosion		Very simple technology for a down-flow co-current adia-batic bed
	Scaling-up	No difficulty: generally lim-ited to batch systems and relatively small sizes	System still poorly known; scale-up should be in steps	Large reactors can be built if liquid distribution is arran-ged carefully

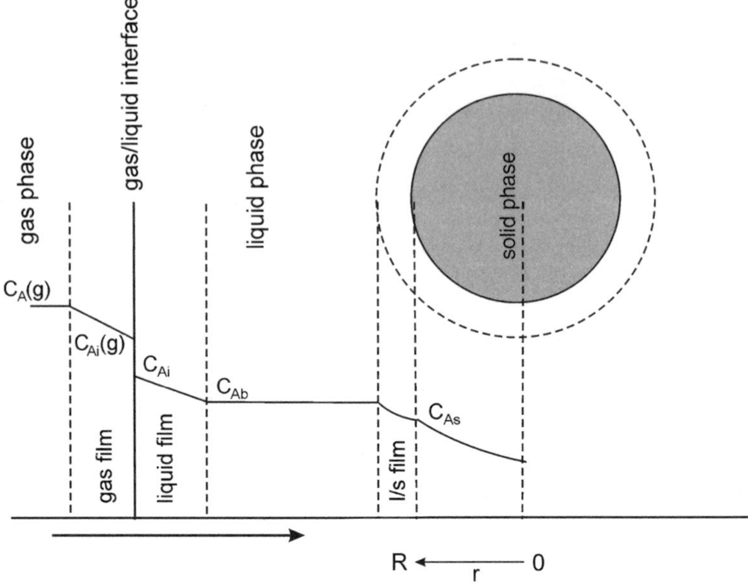

Figure 1.7   Mass transfer and reaction steps in three-phase catalytic reactions.

(iii) Transport from interface to bulk liquid [eqn (1.6)]:

$$-r'_A = k_l a_i \frac{1}{(1 - \varepsilon_b)\rho_c} [c_{Ai} - c_{Ab}] \quad \frac{\text{mol}}{\text{g}_{\text{cat}} \cdot \text{s}} \tag{1.6}$$

where: c_{Ai} = concentration of A in liquid at interface.

(iv) Transport from bulk liquid to external catalyst surface [eqn (1.7)]

$$-r'_A = k_c a_p [c_{Ab} - c_{As}] \quad \frac{\text{mol}}{\text{g}_{\text{cat}} \cdot \text{s}} \tag{1.7}$$

where: a_p = external specific area of pellet; c_{As} = concentration of A at solid–liquid interface; k_c = liquid–solid mass transfer coefficient.

(v) Diffusion and reaction in pellet (assuming second order surface reaction) [eqn (1.8)]

$$-r'_A = \eta \cdot k c_{As} c_{Bs} \quad \frac{\text{mol}}{\text{g}_{\text{cat}} \cdot \text{s}} \tag{1.8}$$

where: c_{Bs} = concentration of B at solid–liquid interface; $k = \eta$ = effectiveness factor defined as the ratio of the reaction rate to the rate in the absence of diffusion.

Combining equations and rearranging gives eqn (1.9), which allows the calculation of the so-called combined mass transfer resistance [expression in the denominator of eqn (1.9)]:

$$-r'_A = \frac{\dfrac{1}{H}}{\dfrac{(1-\varepsilon_b)\rho_c}{Hk_ga_i} + \dfrac{(1-\varepsilon_b)\rho_c}{k_la_i} + \dfrac{1}{k_ca_p} + \dfrac{1}{\eta k c_{Bs}}} \cdot c_A(g) \quad \frac{mol}{g_{cat}s} \qquad (1.9)$$

In modelling the three-phase catalytic reactor, the next stage is to develop material and energy balances. Two approaches are possible: homogeneous or heterogeneous. In the first approach, only the gas phase is considered in the material balance [eqn (1.9) is used]; in the second, one, two or more phases are included in the material balance. For a continuous reactor, these balances are calculated around a differential 'slice' of the reactor (see Figure 1.8) of cross-section S and thickness *dz*. The material and energy balances are coupled *via* a temperature-dependent rate constant (Arrhenius law) [eqn (1.10)]:

$$k = k_0 \exp\left(\frac{-\Delta E_A}{RT}\right) \qquad (1.10)$$

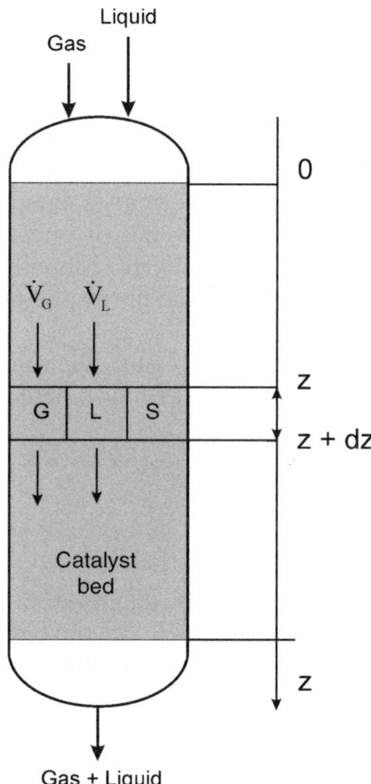

Figure 1.8   Defining the space element for modelling of a fixed bed reactor with co current down-flow.

The correlations necessary to calculate mass transfer coefficients, hold-ups, pressure drops and other model parameters can be found in a number of reviews.[35–36,38,39] In addition, more than 30 Microsoft® Excel simulators developed by the group of Professors Faïçal Larachi and Bernard Grandjean allow the calculations of various model parameters necessary to model trickle bed as well as packed bed reactors. These simulators can be found on the web pages of the Department of Chemical Engineering, Université Laval, Quebec, Canada.[40]

1.3   Scale-up of Conventional Continuous Reactors

One of the major arguments for the transition from conventional large reactor technologies to intensive small reactors is the perceived simplicity of scale-up of the latter. Issues relating to the scale-up of micro- and compact reactors are considered in Section 1.6. Here, we briefly consider the scale-up of conventional reactors, in particular to highlight the differences with the microchemical technologies in the later section.

Within a traditional sequence of process development, the scale-up process is normally performed in several discrete scale steps—from scientific idea to laboratory tests, followed by validation of lab data in a mini-plant, followed by characterisation of fluid dynamics, recycling and stability at a pilot scale and in cold flow models and demonstration units, and only then transferring processes to production-scale units.

The number of steps or the scale-up ratio (defined as a ratio of plant throughputs or as a ratio of linear dimensions of a reactor at two scales) is determined by the confidence in the measured performance characteristics, complexity of the overall process (*e.g.* presence of recycles or integration of heat and mass flows with other units) and whether the reactor is operated in a mass transfer or a kinetic control regime. The latter can be illustrated using the overall reaction rate equations.

For a reaction in a kinetic controlled regime, the change in the concentration of a reactant A for a simple A to B first order scheme can be given by:

$$\frac{C_1}{C_1^0} = e^{-kt} \tag{1.11}$$

where: k = kinetic rate constant.

No terms in eqn (1.11) depend on the length-scale of a reactor. Therefore, such a process, which is entirely under kinetic control, can be scaled up with very large scaling factors. The only limitations to the scale-up ratios are due to:

(i) the level of accuracy of the measured parameters (if the accuracy of, say measured concentrations, is within 10%, a scale-up ratio of 10 may potentially lead to a 100% error); and
(ii) any commercial/business reasoning such as risk *vs.* cost.

In the opposite case [eqn (1.12)], the rate constant k_D is a mass transfer rate constant and therefore necessarily involves distance in its expression. The processes governed by such rate equations require a much more careful scale-up due to the significant dependence of mass transfer steps on the geometry and overall size of the reactors, especially in the case of liquid–solid or three-phase reactions.

$$\frac{C_1}{C_{1,0}} = e^{-k_D a t} \tag{1.12}$$

Thus, for example, if the rate constant in eqn (1.12) represents aeration of a gas–liquid agitated vessel, the actual expression could take the form of the following functional dependence on the reactor geometry:

$$k_D a = f\left(\frac{P}{V}, v_g\right) \tag{1.13}$$

where: a = specific mass transfer area [m^{-1}]; P/V = ratio of agitator power to vessel volume; v_g = superficial gas velocity in the reactor.

In this case in order to achieve the same production yield at different scales (*i.e.* to keep $k_D a$ the same at different reactor scales), it is necessary to maintain the ratios of P/V, v_g and the reactor length/diameter ratio. This example identifies the basic scale-up principle for large reactors—the principle of similarity.

The geometrical similarity (*e.g.* maintaining constant ratios of reactor length to diameter at different scales) is applicable in simple cases when the key mass transfer step is a simple function of a single parameter; for instance, superficial fluid velocity as in the case of gas–liquid bubble columns $[k_L a \propto v \rightarrow \frac{k_L a}{v} = const]$. This is also the case when one considers the relationship between heat removal and surface area of the reactor or the heat transfer elements.

A more complex similarity principle is kinetic similarity, when the preserved quantity is, for example, residence time. In more general terms, the rules of similarity can be represented by the ratios of the dimensionless numbers, which express the key mass or heat transfer mechanisms controlling the reaction production at different scales.[25,41–43]

A particular difficulty of the scaling process is the situation when flow hydrodynamics are the controlling factor. In such cases, simple similarity rules may not produce a satisfactory result. Even in the relatively simple gas–solid reaction system, the complexity of fluid dynamics may not be obvious. Thus, computational fluid dynamics (CFD) simulation of the flow of a gaseous mixture in a chemical vapour deposition reactor reveals a complex flow pattern within the laminar flow range dominated by the buoyancy-induced vortexes, which controls the process performance.[44] Complex fluid flow patterns can also be revealed *via* experimental techniques, such as positron particle tracking, as in the case of gaseous or particulate pneumatic flow through packed beds.[45]

The list of scale-up issues typically addressed in the case of conventional large-scale processes (batch and continuous) include:

- presence of impurities that were not considered/studied in the lab or during the pilot scale tests;
- the fact that explosion limits of mixtures measured in small-scale lab equipment can be narrower than those in the large-scale equipment due to lower rates of heat transfer in the latter;
- shape (agitation, fluid short-circuiting, stagnation zones);
- surface-to-volume ratios, flow patterns;
- construction materials;
- flow stability;
- heat removal;
- wall, edge and end effects.

Microchemical reactor technologies address most of these issues, but especially heat and mass transfer scaling, flow stability, homogeneity of the flow and wall/ edge effects. One most obvious differences between large-scale and micro- chemical reactor technologies is the heat regimes: large-scale reactors are typically operating in an adiabatic regime (apart from multi-tubular reactors such as, for example, steam reforming), whereas microchemical systems can attain a truly isothermal regime, or even be operated with steep temperature gradients or temperature swings.

1.4   Process Intensification: An Overview

In recent years, a significant effort has been devoted to the development of novel techniques and equipment leading to compact, safe, energy-efficient and environmentally friendly, sustainable processes.[15,46,47] These developments focus on 'process intensification'; an approach that has been around for quite some time but has emerged in the past few years as an independent and interesting field of chemical engineering.[15]

Initially, process intensification was defined as a strategy for making a dra- matic reduction in the size of a chemical plant whilst still reaching a given production objective.[48] Such reduction can come from shrinking the size of individual equipment and also from cutting the number of unit operations or apparatuses involved in the entire process. Ramshaw[48] considered volume reduction in the order of 100 or more to be classed as intensification, which is quite a challenging number.

However, this definition describes process intensification exclusively in terms of the reduction in plant or equipment size. More recently, Stankiewicz and Moulijn[15] defined process intensification as:

*" . . . the development of novel apparatuses and techniques that, compared to those commonly used today, are expected to bring **dramatic** improvement in manufacturing and processing, substantially decreasing equipment-size/*

production-capacity ratio, or waste production, and ultimately resulting in cheaper, sustainable technologies".

According to Stankiewicz and Moulijn,[15] the process intensification concept can be divided into two areas:

- process-intensifying equipment (*e.g.* novel reactors, intensive mixing, heat transfer and mass transfer devices);
- process intensifying methods (*e.g.* new or hybrid separations, integration of reaction and separation, heat exchange, phase separation, techniques using alternative energy sources, and new process control methods).

In general, any chemical engineering development that leads to a substantially smaller, cleaner and more energy-efficient technology is process intensification. Examples include:

- novel reactors that provide high surface areas per unit of volume;
- intense mixing devices;
- equipment that performs several unit operations;
- alternative ways of delivering energy to process equipment (*e.g. via* ultrasound, microwaves, light, *etc.*).

These technologies can greatly increase the rate of physical and chemical processes, allowing for high productivity from a smaller volume of a reactor.

Additionally, the important issue is that the strategies for process intensification overlap with those for inherently safer process design.[46,49,50] Reducing the size of a chemical plant generally improves safety by reducing both the quantity of hazardous material that can be released in the case of loss of containment, and the potential energy contained in the equipment (*i.e.* high temperature, high pressure or heat of reaction).

1.4.1 Process-intensifying Equipment

Relevant examples of process-intensifying equipment are briefly described below.

1.4.1.1 Static Mixers

Static mixers are fine examples of process-intensifying equipment (see, for example, www.sulzerchemtech.com). Figure 1.9 shows an example of a static mixer.

Static mixers offer a more size- and energy-efficient method for mixing and can be applied in reaction engineering as static mixer reactors (SMR).[51] Such reactors, which have mixing elements made of heat transfer tubes, can successfully be applied in processes in which simultaneous mixing and intensive heat removal or supply are necessary (*e.g.* nitration, neutralisation reaction, *etc.*). One of the disadvantages of SMRs is their sensitivity to

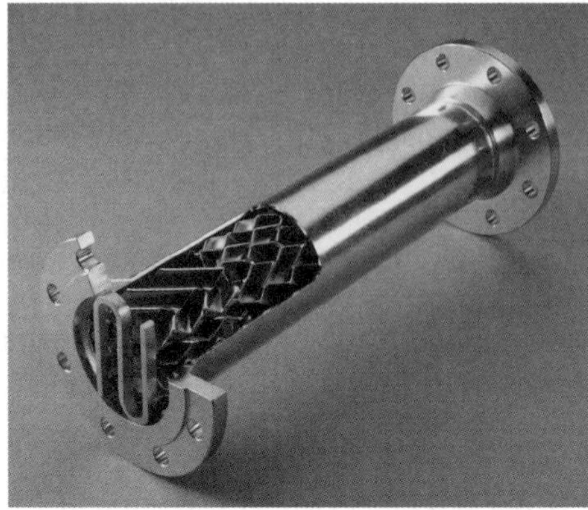

Figure 1.9   Schematic view of a Sulzer static mixer (courtesy of Sulzer and reproduced
with permission).

clogging by solids, making their utility for reactions involving a slurry catalyst
limited.

1.4.1.2 Microreactors

Microreactors, as a novel concept in chemical technology, enable the introduction of new reaction procedures in chemistry, molecular biology and
pharmaceutical chemistry.[1–3,5,10,52] Generally, microreactors are realised as
miniaturised continuous systems or reaction vessels with typical channel or
chamber widths in the range of 10–150 µm. The reduction of characteristic
dimensions, resulting in a reaction zone with a small volume, allows application
of high temperatures or steep reactant concentration gradients, as well as significantly improved process control and enhanced heat management.

Thereby, the unique advantages of microreaction systems are to carry out
chemical reactions in unusual process regimes or under isothermal conditions.
In addition, microreactors enable the production of toxic or explosive chemicals on site or on demand with an inherent safety.

Recently, microreactors have been successfully applied for the synthesis of
vitamin precursors by a homogeneously catalysed two-phase reaction[52] as well
as for partial oxidation of propane to metastable acrolein,[53] ammonia oxidation,[54] water gas shift reaction[55] and many others. A more detailed analysis of
the use of microreactors can be found in Chapter 4.

1.4.1.3 Monolithic Catalysts

The monolith honeycomb structure (defined uniform cross-sectional shape) is
widely used as a catalyst or catalyst support for gas treatment applications
as well as for performing three-phase catalytic reactions.[56–58] For the latter

applications, particular interest has been focused on catalytic reactions such as hydrogenation, oxidation and bioreactions.

The most important features of the monoliths are:

- very low pressure drop in single or two-phase flow;
- high specific surface area;
- high catalytic efficiency due to very short diffusion paths;
- good performance in processes in which selectivity is influenced by mass transfer resistances.

1.4.2 Process-intensifying Methods

As shown in Figure 1.1, most process-intensifying methods fall into three areas:

- integration of reaction and one or more unit operations into so-called multifunctional reactors;
- development of new hybrid separations;
- use of alternative forms and source of energy for processing.

1.4.3 Multifunctional Reactors

The provision of the right amounts of reactants at the reaction site, the establishment and the maintenance of the adequate reaction conditions, and the in-time removal of the reaction products are tasks that are not necessarily solved optimally in standard reactor configurations. Novel designs to improve the interaction of transport and reaction have therefore attracted considerable interest in recent years and have separated into the new field of reaction engineering–multifunctional reactors.[59–63]

The term 'multifunctional reactor' can be defined as reaction equipment in which performance is synergistically enhanced by means of integrating one or more additional process functions.[62] A widely known example of integrating reaction and heat transfer in a multifunctional unit is the reverse flow reactor.[64,65] In such a reactor, the periodic flow reversal allows perfect utilisation of the heat of exothermic reactions by keeping it within the catalyst bed and, after reversion of the flow direction, exploiting stored energy to preheat the cold reactant inlet gases.

Another widely adopted multifunctional reactor is the combined reactor/heat exchanger for fast exothermic chemical reactions, aimed at improving product selectivity and reducing the risk of explosions.[66–68] An example of such a reactor is the diffusion-bonded three-phase compact reactor integrating the functionalities of static mixing, reaction and heat transfer in a single monolithic block; Figure 1.10 shows a schematic representation.[69–71]

An additional functionality that can easily be realised in such reactors is the staged injection of reactants. This is particularly attractive for controlling the selectivity of reactions such as selective oxidations or reductions. Staged

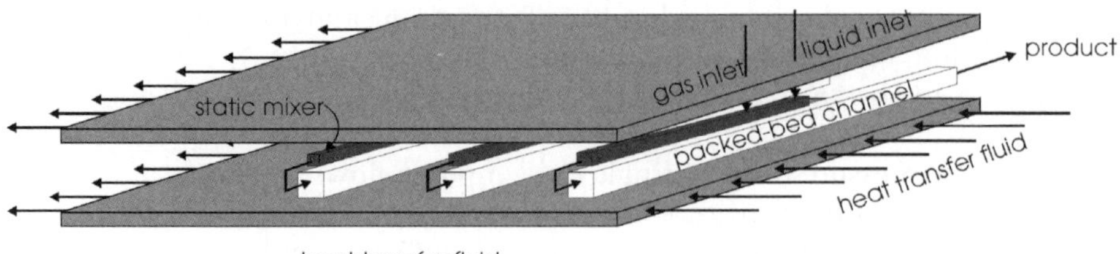

Figure 1.10 Design of a compact multifunctional reactor for selective oxidation of alcohols by molecular oxygen.

Figure 1.11 Principle of staged injection of reactants in a multifunctional reactor demonstrated for a three-phase selective oxidation.

injection can be achieved by, for example, incorporating membranes such as the oxygen sensitive 'chemical valve' membranes,[72] where there is a continuum flux of oxygen through a reactor wall, as shown in Figure 1.11. However, it was subsequently shown that it was sufficient to have a small number of injection points to increase the product yield.[73] The latter is easier to implement in microchemical reactor systems.[71,74]

The concept of combined reactors/heat exchangers can be readily extended to the combination of exothermic and endothermic reactions in a single reactor system.[62,75,76] Such coupled reactions can also exert a favourable influence on the equilibrium, kinetics and selectivity of the synthesis reactions.

Selective removal of products from the middle of an adsorptive reactor is analogous to the withdrawal of a hot gas side stream,[77] and offers an interesting technique for achieving high conversions (*e.g.* for the Claus process[62]). The *in situ* adsorption of water produced by the reaction on zeolite pellets mixed into the catalyst bed can be used to displace the equilibrium in favour of product formation. A similar principle has been proposed for increasing the yield of hydrogen in the steam reforming reaction by means of adsorptive removal of carbon dioxide.[78,79]

Reactive distillation is the other well-known example of integrating reaction and separation in a continuous apparatus.[80] A reactive distillation column is usually split into three sections: (i) the reactive section, in which the reactants are converted into products, and where, by means of distillation, the products are separated out of the reactive zone; the tasks of the rectifying (ii) and stripping (iii) sections depend on the boiling points of the reactants and products.

The advantages of catalytic distillation units, besides the continuous removal of reaction products and higher yields due to the equilibrium shift, consist mainly of reduced energy requirements and lower capital investment.[81] The incorporation of internally finned monoliths into multifunctional reactors for catalytic distillation[82] combines both areas of process intensification.

The other examples of combined reactions and separation processes are:

- reactive extraction;[83,84]
- reactive crystallisation;[85]
- reactive sorption, *i.e.* chromatographic reactors[86–88] and pneumatic transport reactors.[79,89]

In all these processes, the incorporation of a separation unit shifts equilibrium towards product formation.

1.4.4 Membrane Reactors

Membrane reactors have found utility in a broad range of applications including biochemical, chemical, environmental and petrochemical systems.[90,91] A variety of membrane separation processes, the novel characteristics of membrane structures and the geometrical advantages offered by the membrane modules have been employed to enhance and assist reaction schemes to attain higher performance levels compared to conventional approaches.

Membranes perform a wide variety of functions—often more than one function in a given context. They can be employed to:

- introduce/separate/purify reactant(s) and products;
- provide the surface for a reaction;
- provide a structure for the reaction medium;
- retain a specific catalyst.

One of the interesting applications of membranes is the direct bubble free oxygen supply.[92–95]

Control of the reaction pathway is a major concern in reaction engineering. Partial oxidation reactions of hydrocarbons are especially relevant here. Using a membrane to introduce oxygen in a controlled fashion into the reactor can facilitate achievement of the desired reaction conditions (*i.e.* to avoid over-oxidation).

1.4.5 Spinning Disk Reactor

A spinning disk reactor (SDR)[96] is a horizontally circular plate rotating with the speed of 100 to about 6000 rpm. The rotating plate can be cooled or heated with a heat exchange fluid which flows inside the plate. The applied centrifugal force produces thin liquid films on the surface of the rotating disks. Reactants, which are introduced through the centre of the disk, move across the surface forming thin film (the chemical reaction occurs during this step) and are collected on the edge of the disk.

The small dimension of the film (typically $100\,\mu m$) is responsible for very high heat transfer rates between the film and the disk ($14\,kW\,m^{-2}\,K^{-1}$) as well as the high mass transfer between the liquid streams and/or between the film and the gas in the surrounding atmosphere. Additionally, SDRs provide very intense mixing in the thin liquid film and can therefore maintain uniform concentration profiles within a rapidly reacting fluid. The residence time is short and, as a result, may allow the use of higher processing temperatures.

The reactor is characterised by a plug flow. It operates in a safe mode due to the small reactor hold-up and the excellent control of fluid temperature. The apparatus is low fouling and is easy to clean. SDRs have been successfully used to perform fast organic reactions and precipitations,[96] polymerisations[97] or production of nano- and micron-size particles.[96,98]

1.5 Engineering of Multifunctional, Micro- and Compact Reactors

1.5.1 Physics of Flow in Microchannels

One of the key differences between conventional large-scale reactors and most micro- and compact reactor systems is the difference in the flow regimes, or more precisely, the difference in the dominant forces within the flow system.

Most large-scale reactors are dominated by the forces of inertia or large Reynolds numbers (*Re*) and the force of gravity or large Bond numbers (*Bo*), with a smaller importance of viscoelastic interactions and very small effects of surface interactions—unless phenomena at a fluid–solid interface, especially for porous solids, are of significance. Contrary to that, flow in the channels of micro- and compact reactors is largely independent of *Re* numbers, and is dominated by viscoelastic and fluid–surface interactions. For a comprehensive analysis of the physics of fluid flow in microchannels, reader are directed to two very good reviews.[12,99] Table 1.5 lists the various dimensionless numbers useful for describing fluid behaviour in microchannels.

The relative importance of the different forces namely inertia, gravity, surface tension, viscoelastic interactions are illustrated in Figure 1.12. The yellow plane depicts the range of operating conditions and reactor channel sizes where surface interactions dominate over inertia and gravity; the low fluid velocities and small channel diameters are characteristic of microchemical systems. At the micro-level, different multiphase flow structures could be attained in narrow flow channels due to the interplay between effects of pressure, interfacial and intermolecular forces. Thus, pressure-driven behaviour leads to the formation of segmented flow (see Figure 1.13); dominance of capillary forces leads to the

Table 1.5 Dimensionless numbers useful for describing fluid behaviour in microchannels.

Reynolds number inertia *vs.* viscoelastic interactions	$Re = \dfrac{\rho u d}{\mu}$
Peclet number convection *vs.* diffusion	$Pe = \dfrac{l u}{D}$
Bond number gravity *vs.* surface tension	$Bo = \dfrac{(\Delta\rho)g d^2}{\sigma}$
Capillary number viscous *vs.* surface forces	$Ca = \dfrac{\mu u_d}{\sigma}$
Webber number inertia *vs.* surface forces	$We = \dfrac{\rho u_d^2 d}{\sigma}$

where: ρ is fluid density; u is linear velocity [m s^{-1}]; u_d is the linear velocity of the dispersed phase [m s^{-1}]; d is channel/pipe diameter; l is a characteristic length; μ is dynamic viscosity; $(\Delta\rho)$ is the difference in density of immiscible fluids; and σ is surface tension.

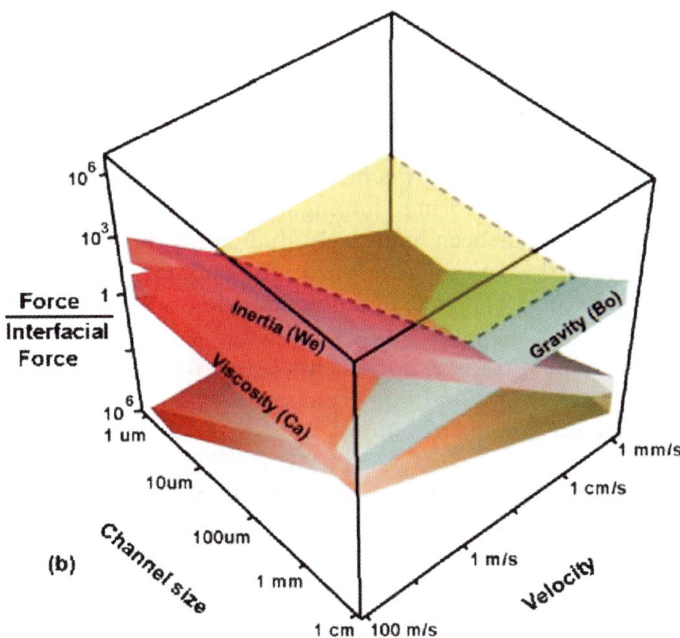

Figure 1.12 Relative importance of different forces as a function of reactor channel diameter (ref. 12, reproduced by permission of The Royal Society of Chemistry).

formation of dispersed small droplets in the so-called 'flow focusing' set up; alternatively, wavy liquid–liquid interfaces could be generated at relatively low velocity differences in the annular flow regime with two immiscible liquids *via* Kelvin–Helmholtz instability.[12]

One of the consequences of the low inertia forces in empty microchannels is the long mixing time, dominated by molecular diffusion. This can be exploited in specific measurement and separation applications, as illustrated in ref. 99. In the case of chemical reactor applications, the long mixing time could be problematic. However, the very high specific surface areas of microreactors, which can be patterned to induce directional changes in flow, allow the design of different types of static mixers such as, for example, the 'chaotic mixer' concept.[100] Even simple static mixer elements, such as periodic baffles, have been shown to be effective in promoting mixing in microchannels.[71] Alternatively, molecular diffusion mixing can be enhanced by further thinning the fluid streams either by using concentric or parallel steams, or in the so-called interdigital laminar micromixers commercialised by the Institut für Mikrotechnik Mainz GmbH (IMM).[1]

1.5.2 Principles of Multiphase Contacting in Micro- and Compact Reactors

The flexibility in design, access to a wide range of construction materials and the ability to utilise capillary forces and surface–fluid interactions result in a large number of possible arrangements for phase contacting in compact and microreactors.[101]

Dispersive contacting is typically achieved in slug (Taylor) flow, foam flow or bubble flow, as well as in the form of impinging jet reactors (see Figure 1.13). In particular, Taylor flow has been intensively studied for the cases of multiphase monolith, membrane and microreactors; see, for example, refs. 56,95,102–109.

The attractive features of segmented flow include:

- relatively high mass transfer coefficients between capillary wall and liquid phase, and between gas and liquid phase;[95,104]
- narrow residence time distribution.[106]

In particular the latter, in combination with enhanced mixing in meandering rather than straight channels, has been exploited for the synthesis of particles with a narrow particle size distribution *via* a sol–gel route.[106] The enhanced mass transfer is sufficient to produce a marked increase in the observed reaction rates of common chemical reactions.[110]

Bubble flow can be efficiently realised in micro-packed beds in the case of three-phase reactions. Such a system has been proposed for the selective oxidation of alcohols in a compact multifunctional reactor.[69,71] An idealised two-dimensional representation of bubble flow is shown in Figure 1.13: microspherical supports with a uniform particle size provide a convenient way to pack millimetre-scale reactor channels and also result in high Péclet numbers (*Pe*) (*i.e.* behaviour close to plug flow), and good hydrodynamic stability (low and stable pressure drop), see Figure 1.14.[70,71] Conversely, randomly shaped catalyst support particles result in larger pressure drops and unstable hydrodynamics. In both cases (segmented

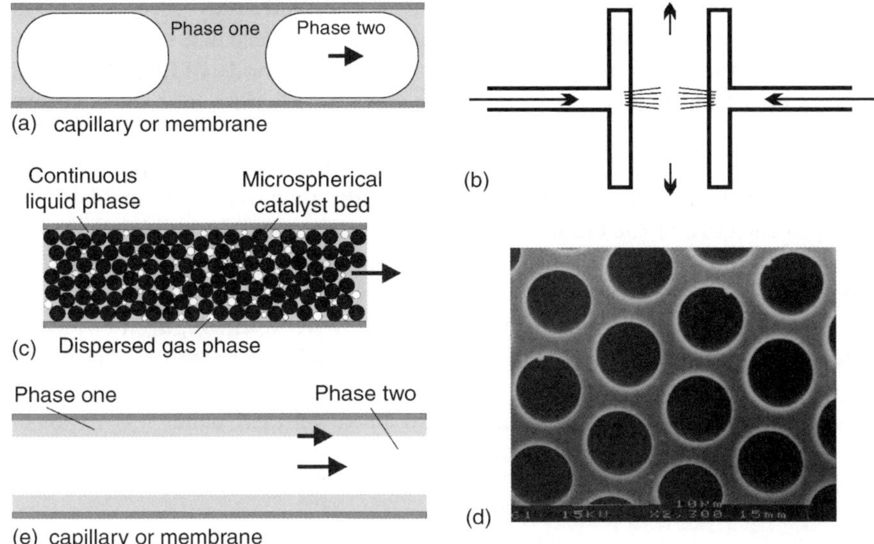

Figure 1.13 Different types of dispersive and non-dispersive fluid contacting in micro- and compact reactors. (i) Dispersive contacting in micro- and compact reaction systems: (a) a scheme of slug or Taylor flow; (b) impinging jets; (c) bubble flow (ii) Non-dispersive contacting in micro- and compact reaction systems: (d) a structured glass membrane for non-dispersive phase contacting; (e) a scheme of annular flow.

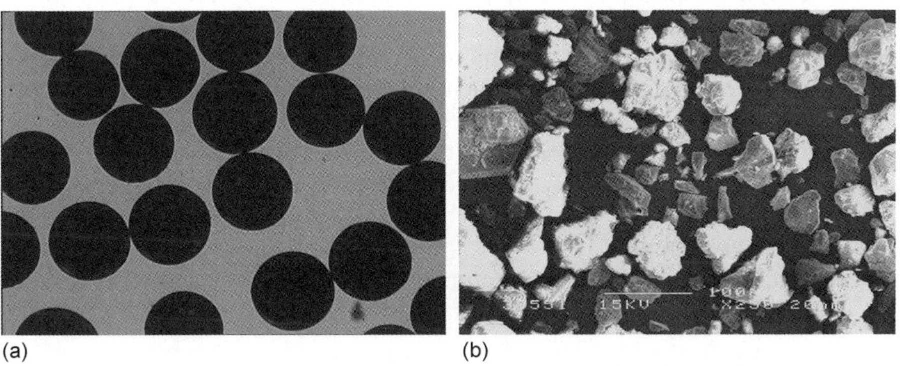

Figure 1.14 Scanning electron microscopy (SEM) images of (a) Novacarb™ meso-porous synthetic carbon catalyst support (mean particle size *ca.* 160 µm) and (b) Ru(III)/Al$_2$O$_3$ catalyst both used in a multifunctional three-phase compact reactor.

and dispersed bubble flows), the continuous phase is characterised by a laminar flow regime, leading to a relatively slow mixing.

A very different approach to dispersive multiphase reactors, which can be realised also in microchemical technology, is impinging jet reactors.[111–115]

Collision of high-velocity fluids streams in a confined space creates high turbulence regimes, promoting mixing and heat transfer. Because of the enhanced heat transfer, there is a significant interest and a large body of literature on the applications of impinging jets in heat transfer and burning, but these are outside the scope of this chapter.

Non-dispersive contacting can be arranged in an annular or pipe flow,[116,117] with the thin liquid film flowing along the walls of narrow channels and gas flowing in the core of the channel, or with the much lower fluid linear velocities in the falling film reactors[118] and in membrane contactors.[119,120] A comparison of the same reaction performed in dispersed Taylor flow and in annular flow regimes showed an appreciable increase in the reactor throughput.[117] Microfabricated membranes, such as the glass plate shown in Figure 1.13, allow very tight control over mass transfer across the interface due to uniformity of pores. However, membrane contactors may also be developed using less expensive techniques, for example, *via* the use of thin porous plate membranes[120] or hollow fibre membrane modules.[90]

1.5.3 Heterogeneous Catalyst Design for Micro- and Compact Reactors

The majority of well-studied reactions in micro- and compact reactors are either uncatalysed stoichiometric or homogeneously catalysed single and two-phase reactions. These do not represent particular challenges for micro- or compact reactors, apart from those cases when more viscous fluids are being used (*e.g.* ionic liquids) or when precipitation of an insoluble product/by-product is possible.

Heterogenisation of catalysts in microchemical systems on another hand is a challenge. There are four key approaches developed in the literature:

- coated wall reactors with a catalytic 'washcoat';[121–124]
- grafted or non-covalently bonded molecular catalysts on the walls of empty channels;
- micro-packed beds;[69,125]
- channels filled with a structured packing, *e.g.* polymeric porous materials.[126,127]

1.5.3.1 Coated Wall Reactors

Coated wall microcapillaries and structured reactors are widely investigated for petrochemical-type reactions (*e.g.* reforming, shift, *etc.*) as well as fine chemistry applications.

With such reactors, there is an inevitable compromise between the considerable advantages of microreactors, namely:

- high rates of mass transfer;
- good control over flow distribution hence selectivity;

- high rates of heat transfer;
- possibility for staged injection of reactants;
- possibility for using magnetic and electric fields;
- flow manipulation by controlling interfacial tensions;
- use of *in situ* analytics *etc*,

And a rather long list of drawbacks, including most obvious:

- difficulty in creating perfectly adhered catalytic layers on microstructures;
- high cost of manufacturing;
- single use (difficult to remake catalytic layers after deactivation);
- cost of scaling-up (or scaling-out).

However, there is significant potential for developing such reactor systems if the material science problems (adherence) and cost of manufacturing (see below) could be resolved. Several recent studies have demonstrated effective synthesis of coated wall porous supports and catalysts with apparent high stability and catalyst effectiveness. Thus, Pd nanoparticle catalysts or bimetallic catalysts deposited into a porous titania layer in a fused capillary microreactor is described in ref. 122. Another class of oxide supports that is frequently used in catalysis is mesoporous silica. It has been demonstrated in a number of studies (see, for example, ref. 128) that the open mesoporous structure of a support is beneficial for mass transfer and efficiency of the supported catalysts or chemisorbents. Preparation of a templated mesoporous silica inside microcapillaries was described recently in ref. 129.

1.5.3.2　Grafting

Grafting different inorganic or organic structures to attach active catalyst sites, ion exchange or other bonding sites is a well-investigated technique and we will not spend much time on discussing its applications for microreactors. However, two topics are worth mentioning. Recently, particular interest has emerged in the use of porous silicon as a matrix for anchoring active groups, or as a reactive template for the synthesis of metal nanoparticles.[130] Porous silicon can also be formed inside microreactor channels and used as a template for grafting active groups or depositing catalysts. The interested reader is directed to recent papers on this topic.[131–133] The second topic is flow manipulation by using responsive polymers deposited onto reactor walls.[134] This development offers tremendous opportunities for active fluid manipulation in compact and microreactors using temperature, electric potential, light or specific chemical traces as controlling signals.

1.5.3.3　Micro-Packed Bed

Of the four identified types of heterogeneous compact and microreactor systems, the use of micro-packed beds is arguably the most versatile approach since it utilises the wealth of knowledge in the catalysis community on the design of porous heterogeneous catalysts.

A micro-packed bed reactor for liquid phase reactions is described in ref. 125. The device contains catalyst filling channels, as well as internal filters, to prevent loss of catalyst particles with the flow of reactants. A larger diameter capillary reactor is described for the direct synthesis of hydrogen peroxide.[135] In this work, a 1/16 inch tubing with 0.765 mm internal diameter (ID) was used as a reactor. The flow regime was not well characterised and is described as broken Taylor flow (see definitions of multiphase contacting in microreactors in Section 1.5.2).

A millimetre-scale compact multifunctional reactor with a microspherical packed bed is described in detail in a series of papers.[69–71] The microspherical carbon support was used to enable easy catalyst channels loading by free-flowing solid particles, exploiting the flexibility of the design of the phenolic resin based structured micro–meso porous carbons.[136,137] The same reactor packed with randomly shaped alumina or titanate nanotube supports exhibited much higher pressure drops and worse hydrodynamic stability than in the case of the spherical monodispersed particles.

A combination of coated wall and packed bed concepts was proposed on the basis of a microstructured reactor environment.[121] In this case, the high surface/volume ratio required to achieve a high catalytic area is attained by producing structured micro-pillars filling the reactor volume. The structured packing results in both high specific area and low pressure drop—the same effect which is achieved by packing membranes into bundles of narrow capillaries (hollow fibres) or by depositing catalysts onto walls of monoliths with high cell density. The micro-pillar geometry of microreactors is highly efficient, but is expensive to produce.

1.5.3.4   *Channels Filled with a Structured Packing*

An alternative packing for microreactors is a functionalised porous polymer matrix either as a porous polymeric monolith or deposited onto porous glass.[126–127,138,139] Polymeric support offers broad functionality for attaching catalytic functions, which is particularly attractive for anchoring organometallic catalysts.[126,127] Polymers also offer good flexibility for the manufacture of different flow patterns through the monolith and the resultant framework can be used to deposit the catalyst containing layers, as for example in the case of supported ionic liquids.[139]

1.5.4   Fabrication of Micro- and Compact Reactors

One of the biggest challenges for rapid implementation of micro- and compact reactor technologies is the use of expensive fabrication methods. Here we describe only several key techniques. For earlier reviews on the topic the reader is directed towards an excellent book.[1]

1.5.4.1 Diffusion Bonding and Mechanical Assembly of Metal Plates

The chemical etching of polished thin metal plates to produce groves or through channels and holes, followed by assembly of the pack of plates and their diffusion bonding in a high temperature press/furnace, results in a three-dimensional (3D) structured metal monolith with complex internal structures. This process has been perfected by Chart Heat Exchangers (www.chartindustries.com) and Heatric (www.heatric.com), which use either grooved plates or through channels in plates (shims) to assemble their structures. Heatric produces very large compact heat exchangers based on this technology, but also produces a number of custom-designed research modules, including a pilot plant module for our group.[140]

This technology can produce complex internal structures in an efficient 3D geometry and is suitable for the development of micro packed bed or coated wall micro- and compact reactors. Because the resultant structures are effectively metal monoliths, the diffusion-bonded reactors can be developed for high-pressure applications. The downsides of the technology are the high cost of manufacturing, reliance on chemical etching which produces a significant amount of chemical waste, and long lead times for developing protocols for extending the range of metals which can be used in the diffusion-bonding process.

A simplified method of assembling individually machined or etched plates is to apply pressure to a stack of plates by bolts or sealing the plates *via* external welding. Examples of such reactors are the corrugated metal plate reactor described in an early patent,[141] a metal plate reactor/heat exchanger for fluorination reactions described in ref. 116 or a metal foil assembly reactor for compact fuel processing.[142] However, such structures often suffer from internal leaks and buckling of plates, and hence poor flow distribution or even blockages.

1.5.4.2 Etched/Micromachined Silicon, Plastic and Glass

There are many descriptions of different microreactor designs in silicon or glass. Examples include photochemical reactors etched in silicon wafers,[143] or either etched or machined, and then anodically bonded glass plates.[144] This platform is the most widely explored and a number of commercial providers and research institutes offer microreactors etched or machined in silicon, glass and polymers. Most of such devices are open channel structures suitable as residence-time modules or mixing modules for lab-scale flow chemistry studies. However, more complex structures can be developed. For example, a highly regular low pressure drop catalyst support fabricated in a silicon microreactor is described in ref. 121.

1.5.4.3 Moulded Polymers, Ceramics; Rapid Prototyping

Significant advantages over etching, machining and diffusion bonding exist in the polymer and metal fabrication methods based on moulding, stamping and rapid prototyping. Early adoption of microfluidics in medicinal and bio-chemical applications was due to access to cheap devices on the basis of polydimethylsiloxane (PDMS), which can be cast into moulds or formed using rapid prototyping at very low cost and with sufficient precision in the device's features; see ref. 145 for a good review on PDMS applications in microreactors. Alternative polymers could be used to increase the solvent tolerance of the devices.[146] Moulding and casting techniques can also be used to produce microreactor elements in ceramics.[147,148]

Cheap fabrication methods as moulding and rolling can also be used to form channels of holes in thin metal films, which could later be assembled into microreactors. Rapid prototyping could also be extended to metals *via* the so-called selective laser melting (SLM) technique, which allows the manufacturing of 3D monolith structures from fine metal powders.

1.6   Scale-up of Micro- and Compact Reactors

The issues of scale-up in microchemical reactor systems differ radically from the typical scale-up process of conventional large-scale reactors. Thus, heat transfer management to avoid hot spots or mass transfer problems induced by, for example, flow inhomogeneity are of little importance. However, new challenges have appeared. One of the key ideas behind microchemical tech-nology is the ability to maintain the reaction conditions/features of the small-scale reactors at the scale of industrial throughput, which is believed to be attainable by the numbering up of small reactors into large parallel reactor systems, as shown conceptually in Figure 1.15.

New types of scale-up/scale-out issues have emerged for microchemical reactor systems including:

- blockage by particulates of narrow reaction and heat transfer channels;
- inhomogeneous distribution of flow between parallel reaction channels;
- cost of manufacturing reactors;
- precision of flow control, pulse-free flow;

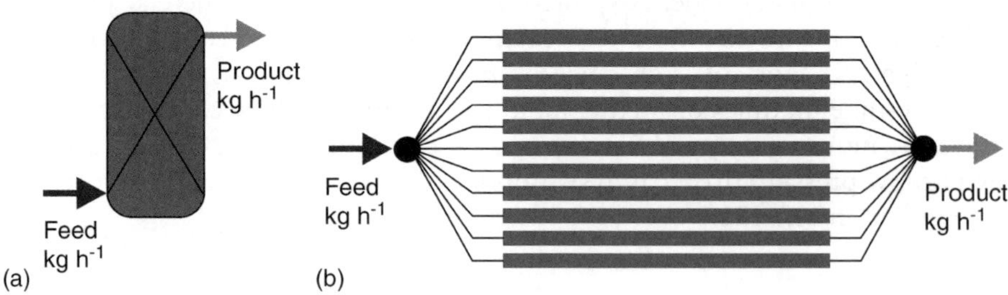

Figure 1.15   Conceptual differences between parallel system of micro- or compact reactors (b) and the traditional scaled-up reactor with a similar throughput (a).

- pressure drop in narrow reaction and heat transfer channels;
- control of large parallel reaction systems.

Below we briefly discuss some of the specific issues of the scale-up/scale-out of compact and microreactor systems.

1.6.1 Blockage of Microreactors

The problem of solids in microreactors is widely appreciated. Solids originate from:

- industrial quality solutions/reagents;
- debris from other equipment and infrastructure;
- the synthesis of low solubility by-products;
- the synthesis of solids.

Thus, in our studies of selective oxidation of benzyl alcohol in benzaldehyde,[69–71] slow formation of insoluble benzoic acid resulted in (i) reversible deactivation of catalysts and (ii) increase in the pressure drop across micro-packed beds. However, it was shown that, by careful design of wetting of

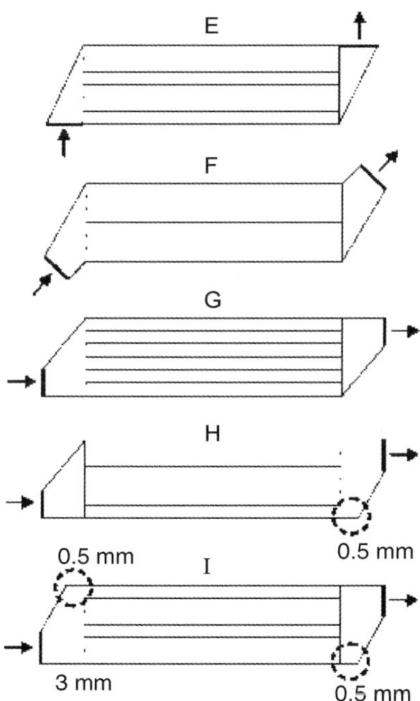

Figure 1.16   Optimisation of the fluid inlets and outlets, and headers for multi-channel parallel reactor plates (reproduced with permission from ref. 151).

microchannels, it is possible to avoid clogging even in reactions involving the synthesis of solids.[149] This particular approach is reliant on segmented (Taylor) flow behaviour, in which the dispersed phase does not come into direct contact with the capillary wall. In this case, the solids are confined to the dispersed phase and thus do not build-up on the reactor walls. It is an efficient method of synthesis of the uniform sized micro- or nanoparticles.

1.6.2 Flow Distribution in Multiple Parallel Channels

One of the key problems in scaling-out of microchemical reactor systems based on the concept of multiple parallel reaction channels is the establishment of a uniform flow/reactants distribution among all the parallel reactor channels. Inability to reach this goal completely undermines the purpose of the whole concept, since good control over reaction selectivity cannot then be achieved. Therefore, considerable attention has been paid to the design of flow headers for connecting different types of compact and microreactors to inlet and outlet flow pipes.

The design of flow distributors for parallel channels in plate reactors or heat exchangers requires full computational fluid dynamics (CFD) simulation to

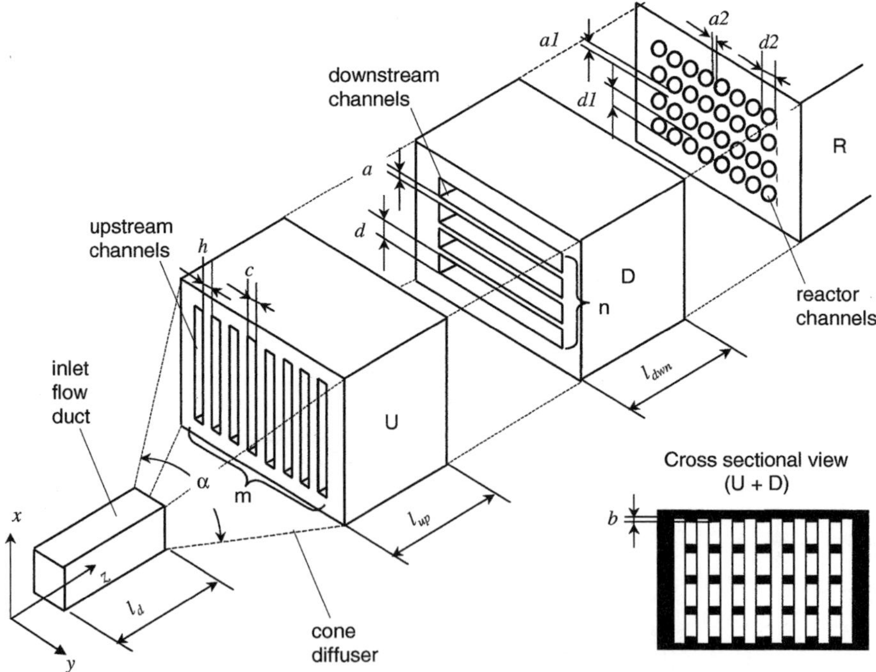

Figure 1.17   Design of a header for connecting an inlet feed to a multi-tube reactor (reproduced with permission from ref. 152).

Figure 1.18   Design of a gas inlet with a pressure drop section for equal distribution of gas into individual channels.

avoid the imbalance of flow, and hence sub-optimal selectivity/performance of microreactors.[150] An example of CFD design of flow distributor geometry for fast flow in a short contact time in a microreactor is given in ref. 151. The geometries used for optimisation of the inlet and outlet fluid headers, reproduced in Figure 1.16, are not necessarily apparent, and the optimum geometry cannot be arrived at without CFD.

A very useful concept for flow distribution in 3D microreactors is the design of a so-called 'thick-wall' screen header,[152] as illustrated in Figure 1.17. Uniformity of flow across all channels of the reactor module (R) is achieved by splitting the flow into thinner slices in two modules (U and D), with perpendicular orientation of thick-wall screens. The effect of the design variables (sizes of screens: h, c, a, d, l_{up} and l_{down}) on the screen efficiency have been studied carefully and design equations for the optimal screen geometry developed.[153]

Another fluid distribution problem is when a single gas (or more generally a condensable fluid) stream must be separated into multiple reaction or mixing channels. The downstream pressure drop can have a significant influence on the behaviour of the fluid in the entrance region and on the distribution between the channels. In this case, a pressure-drop section and percolating of the channel diameter are frequently employed for the purpose of flow equilibration; Figure 1.18 shows an example of such a design.

1.7   Concluding Remarks

Continuous chemical processes characterised by highly integrated material and energy flows are well-established in traditional petrochemical industries.

They are the workhorses of current chemical technologies, which were the foundation of the tremendous success of industrialised economies.

The current drive towards developing more sustainable chemical processes has drawn attention towards creating a new range of continuous processes, suitable for smaller scale manufacturing in speciality chemicals, pharmaceuticals, nutraceuticals, food and fragrance industries, *etc.* Some of these processes are more widely known as 'flow-chemistry'. It is a new field, but there are already some remarkable success stories about commercial realisations of compact and microreactors, and other intensified reactor technologies.[1,3,68,154,155] There are also many new avenues for development of these technologies, which should lead to significant technological advances (*e.g.* in photochemistry),[143,144,156–158] integrating reactors with sensors and smart elements,[159] and integration of reactor technology with design of nanomaterials.[160]

The engineering of flow reactors for new applications cannot rely on the traditional rule book. Not only is the physics of flow in narrow channels different to that in large channels (see Section 1.5.1), but many manufacturing principles and manufacturing technologies are also rather different. Thus, the conventional pressure regulations for large vessels do not work for microreactors and attempts to use them result in disaster. For example, a combined reactor/heat exchanger with an inbuilt highly efficient micro-heat exchanger, assembled into a metal body with flanges designed according to a conventional pressure vessel handbook would simply not work, since the thermal mass of the device would kill any advantages of the advanced device architecture.

At the same time, advances in manufacturing methods (*e.g.* the fast evolution of the rapid prototyping techniques) will considerably reduce the development time of novel flow reactors and also reduce their cost. There is a strong argument for close integration of the development of novel chemistry alongside the development of novel processing technologies in a concerted manner, and also involving the full arsenal of reaction engineering: computational fluid dynamics, CAD design, flow visualisation, reactor simulation, *in situ* analytics, *etc.* Only in such a manner will a true breakthrough in these technologies be developed.

1.8   Symbols/Nomenclature

a_i gas–liquid interfacial area / volume of bed [$m^2 \, m^{-3}$]

a_p external specific area of pellet [$m^2 \, gcat^{-1}$]

$c_A(g)$ bulk gas phase concentration of A [$kmol \, m^{-3}$]

$c_{Ai}(g)$ interfacial concentration of A [$kmol \, m^{-3}$]

c_{Ai} concentration of A in liquid at interface [$kmol \, m^{-3}$]

c_{Ab} bulk liquid concentration of A [$kmol \, m^{-3}$]

c_{As} concentration of A at solid–liquid interface [$kmol \, m^{-3}$]

d diameter, channel diameter, hydraulic diameter [m]

E enhancement factor

H $\quad$ Henry's constant

k $\quad$ specific reaction constant [m^3 of liquid mol^{-1} $gcat^{-1}$ s^{-1}]

k_g $\quad$ gas phase mass transfer coefficient [m s^{-1}]

k_l $\quad$ liquid phase mass transfer coefficient [m s^{-1}]

k_c $\quad$ liquid–solid mass transfer coefficient [m s^{-1}]

k_{Al} $\quad$ mass transfer coefficient in the liquid phase [m s^{-1}]

k_{Ag} $\quad$ mass transfer coefficient in the gas phase [m s^{-1}]

l $\quad$ length [m]

p_A $\quad$ partial pressure of A in the gas phase [MPa]

u $\quad$ linear fluid velocity, superficial velocity [m s^{-1}]

u_d $\quad$ linear characteristic velocity of the dispersed phase [m s^{-1}]

z $\quad$ dimensionless length

δ_l $\quad$ thickness of the laminar layer [m]

ε_b $\quad$ bed porosity (gas + liquid) [–]

μ $\quad$ dynamic viscosity [N s m^{-2}]

η $\quad$ effectiveness factor

ρ_c $\quad$ density of catalyst pellet [kg m^{-3}]

ρ $\quad$ density [kg m^{-3}]

$(\Delta\rho)$ $\quad$ density difference between immiscible fluids [kg m^{-3}]

σ $\quad$ surface tension [N m^{-1}]

References

1. W. Ehrfeld, V. Hessel and H. Löwe, *Microreactors,* Wiley-VCH, Weinheim, 2000.
2. K. Jähnisch, V. Hessel, H. Löwe and M. Baerns, *Angew. Chem. Int. Ed.,* 2004, **43**, 406.
3. K. F. Jensen, *Chem. Eng. Sci.,* 2001, **56**, 293.
4. J. Kobayashi, Y. Mori and S. Kobayashi, *Chem. Asian J.,* 2006, **1–2**, 22.
5. S. Taghavi-Moghadam, A. Kleemann and K. G. Golbig, *Org. Process Res. Dev.,* 2001, **5**, 652.
6. T. L. LaPorte and C. Wang, *Curr. Opin. Drug Discov. Dev.,* 2007, **10**, 738.
7. T. Laird, *Org. Process Res. Dev.,* 2007, **11**, 927.
8. L. Proctor, *Chemical Technology,* 84.
9. R. C. Wheeler, O. Benali, M. Deal, E. Farrant, S. J. F. MacDonald and B. H. Warrington, *Org. Process Res. Dev.,* 2007, **11**, 704.
10. D. M. Roberge, B. Zimmermann, F. Rainone, M. Gottsponer, M. Eyholzer and N. Kockmann, *Org. Process Res. Dev.,* 2008, **12**, 905.
11. M. Baumann, I. R. Baxendale, S. V. Ley, N. Nikbin, C. D. Smith and J. P. Tierney, *Org. Biomol. Chem.,* 2008, **6**, 1577.
12. A. Günther and K. F. Jensen, *Lab Chip,* 2006, **6**, 1487.
13. A. Warmington and C. Challener, *Speciality Chemicals Magazine,* 2008, **40**.
14. N. G. Anderson, *Org. Process Res. Dev.,* 2001, **5**, 613.

15. A. I. Stankiewicz and J. A. Moulijn, *Chem. Eng. Prog.*, 2000, **96**, 22.
16. A. Stankiewicz, *Chem. Eng. Sci.*, 2001, **56**, 359.
17. M. Krishnan, V. Namasivayam, R. Lin, R. Pal and M. A. Burns, *Curr. Opin. Biotechnol.*, 2001, **12**, 92.
18. M. Toner and D. Irimia, *Annu. Rev. Biomed. Eng.*, 2005, **7**, 77.
19. M.-O. Coppens, *Ind. Eng. Chem. Res.*, 2005, **44**, 5011.
20. C. Situma, M. Hashimoto and S. A. Soper, *Biomol. Eng.*, 2006, **23**, 213.
21. J. C. Giddings, *Science*, 1993, **260**, 1456.
22. C. O. Kappe, *Chem. Soc. Rev.*, 2008, **37**, 1127.
23. C. A. Kayode, *Ludwig's Applied Process Design for Chemical and Petrochemical Plants,* Elsevier, Amsterdam and Boston, 2007.
24. P. H. Spitz, *Petrochemicals: the Rise of an Industry,* Wiley, New York and Chichester, 1988.
25. P. Trambouze, H. v. Landeghem and J. P. Wauquier, *Chemical Reactors: Design, Engineering, Operation,* Gulf Publishing, Houston, 1988.
26. L. K. Doraiswamy and D. B. Kulkarni, in *Chemical Reaction and Reactor Engineering,* ed. J. J. Carbery and A. Varma, M. Dekker, New York, 1987.
27. R. E. Rosensweig, US Pat., 4 115 927, 1978.
28. R. E. Rosensweig, *Science*, 1979, **204**, 57.
29. S. Carra and M. Morbidelli, in *Chemical Reaction and Reactor Engineering*, ed. J. J. Carberry and A. Varma, M. Dekker, New York, 1987.
30. M. Baerns, H. Hoffman and A. Renken, *Chemische Reaktionstechnik,* Thieme, Stuttgart, 1999.
31. Y. T. Shah and M. M. Sharma, in *Chemical Reaction and Reactor Engineering*, ed. J. J. Carbery and A. Varma, M. Dekker, New York, 1987.
32. P. A. Ramachandran and R. V. Chaudhari, *Three-Phase Catalytic Reactors,* Gordon & Breach Science Publishers, 1983.
33. J. Hagen, *Industrial Catalysis,* Wiley-VCH, Weinheim, 1999.
34. A. I. van Diepen and J. A. Moulijn, in *Fine Chemicals through Heterogeneous Catalysis*, ed. R. A. Sheldon and H. v. Bekkum, Wiley-VCH, Weinheim, 2001.
35. M. P. Dudukovic, F. Larachi and P. L. Mills, *Chem. Eng. Sci.*, 1999, **54**, 1975.
36. M. P. Dudukovic, F. Larachi and P. L. Mills, *Catal. Rev.*, 2002, **44**, 123.
37. H. S. Fogler, *Elements of Chemical Reaction Engineering,* Prentice-Hall International, Upper Saddle River, NJ, 1999.
38. A. Gianetto and V. Specchia, *Chem. Eng. Sci.*, 1992, **47**, 3197.
39. M. H. Al-Dahhan, F. Larachi, M. P. Dudukovic and A. Laurent, *Ind. Eng. Chem. Res.*, 1997, **36**, 3292.
40. www.gch.ulaval.ca/flarachi/ and www.gch.ulaval.ca/bgrandjean/ [accessed June 2009].
41. S. T. Sie and R. Krishna, *Rev. Chem. Eng.*, 1998, **14**, 47–88, 109–157, 158–202, 203.

42. A. Bisio, *Scale-up of Chemical Processes: Conversion from Laboratory Scale Tests to Successful Commercial Size Design,* Wiley, New York and Chichester, 1985.
43. G. Caygill, M. Zanfir and A. Gavriilidis, *Org. Process Res. Dev.*, 2006, **10**, 539.
44. Ł. J. Sytniewski, A. A. Lapkin, S. Stepanov and W. N. Wang, *J. Cryst. Growth*, 2008, **310**, 3358.
45. Y. Ding, Z. Wang, D. Wen, M. Ghadiri, X. Fan and D. Parker, *Chem. Eng. Sci.*, 2005, **60**, 5231.
46. D. C. Hendershot, *Chem. Eng. Processing*, 2000, **96**, 35.
47. G. E. Keller and P. F. Bryan, *Chem. Eng. Processing*, 2000, **96**, 41.
48. C. Ramshaw, *1st International Conference on Process Intensification for the Chemical Industry*, Mechanical Engineering Publications, London, 1995, BHR Group Limited Conference Series Publication No. 18.
49. R. S. Benson and J. W. Ponton, *Trans. IChemE*, 1993, **71A**, 160.
50. D. C. Hendershot, *J. Loss Prevent. Proc. Ind.*, 1997, **10**, 151.
51. C. Brechtelsbauer and F. Ricard, *Org. Process Res. Dev.*, 2001, **5**, 646.
52. H. Loewe and W. Ehrfeld, *Electrochimica Acta*, 1999, **44**, 3679.
53. D. Hoenicke and G. Wiessmeier, in *Microsystem Technology for Chemical and Biological Microreactors*, DECHEMA, 1996, vol. 132, p. 93.
54. R. Srinivasan, I.-M. Hsing, P. E. Berger, K. F. Jensen, S. L. Firebaugh, M. A. Schmidt, M. P. Harold, J. J. Lerou and J. F. Ryley, *AIChE J.*, 1997, **43**, 3059.
55. A. Y. Tonkovich, J. L. Zilka, M. J. LaMont, Y. Wang and R. S. Wegeng, *Chem. Eng. Sci.*, 1999, **54**, 2947.
56. S. Irandoust and B. Anderson, *Catal. Rev.-Sci. Eng.*, 1988, **30**, 341.
57. A. Cybulski and J. A. Mulijn, *Catal. Rev.-Sci. Eng.*, 1994, **36**, 170.
58. A. Cybulski and J. A. Mulijn, *Structured Catalysts and Reactors,* Marcel Dekker, New York, 1998.
59. D. W. Agar and W. Ruppel, *Chem.-Ing. Tech.*, 1988, **60**, 731.
60. K. R. Westerterp, *Chem. Eng. Sci.*, 1992, **47**, 2195.
61. U. Hoffmann and K. Sundmacher, *Chem.-Ing. Tech.*, 1997, **69**, 613.
62. D. W. Agar, *Chem. Eng. Sci.*, 1999, **54**, 1299.
63. G. J. Harmsen and L. A. Chewter, *Chem. Eng. Sci.*, 1999, **54**, 1541.
64. G. K. Boreskov, Y. S. Matros and O. V. Kiselev, *Kinet. Katal.*, 1979, **20**, 773.
65. Y. S. Matros and G. A. Bunimovich, *Catal. Rev.-Sci. Eng.*, 1996, **38**, 1.
66. C. H. Phillips, G. Lauschke and H. Peerhossaini, *Appl. Thermal Eng.*, 1997, **17**, 809.
67. E. V. Rebrov, M. H. J. M. d. Croon and J. C. Schouten, *Catal. Today*, 2001, **69**, 183.
68. Z. Anxionnaz, M. Cabassud, C. Gourdon and P. Tochonb, *Chem. Eng. Processing*, 2008, **47**, 2029.
69. D. V. Bavykin, A. A. Lapkin, S. T. Kolaczkowski and P. K. Plucinski, *Appl. Catal. A: Gen.*, 2005, **288**, 165.
70. P. K. Plucinski, D. V. Bavykin, S. T. Kolaczkowski and A. A. Lapkin, *Catal. Today*, 2005, **105**, 479.

71. P. K. Plucinski, D. V. Bavykin, S. T. Kolaczkowski and A. A. Lapkin, *Ind. Eng. Chem. Res.*, 2005, **44**, 9683.

72. A. Julbe, D. Farrusseng and C. Guizard, *Sep. Purif. Technol.*, 2001, **25**, 11.

73. S. Thomas, S. Pushpavanam and A. Seidel-Morgenstern, *Ind. Eng. Chem. Res.*, 2004, **43**, 969.

74. L. Guillou, S. Paul and V. L. Courtois, *Chem. Eng. J.*, 2008, **136**, 66.

75. R. Charlesworth, A. Gouugh and C. Ramshaw, *4th UK National Conference on Heat Transfer,* Mechanical Engineering Publications, London, 1995.

76. F. van Looij, Utrecht University, 1994.

77. U. Nieken, *Abluftreinigung in katalytischen Festbettreaktoren bei periodischer Strömungsumkehr,* VDI-Vortschritt-Berichte, Düsseldorf, 1993.

78. Y. Ding and E. Alpay, *Chem. Eng. Sci.*, 2000, **55**, 3929.

79. G. C. Koumpouras, E. Alpay and F. Stepanek, *Chem. Eng. Sci.*, 2007, **62**, 2833.

80. J. D. Seader and E. J. Henley, *Separation Process Principles,* John Wiley, New York, 1998.

81. W. P. Stadig, *Chem. Process.*, 1987, **50**, 27.

82. P. J. M. Lebens, F. Kapteijn, S. T. Sie and J. A. Moulijn, *Chem. Eng. Sci.*, 1999, **54**, 1359.

83. M. Minotti, M. F. Doherty and M. F. Malone, *Ind. Eng. Chem. Res.*, 1998, **37**, 4748.

84. K. D. Samant and K. M. Ng, *Chem. Eng. Technol.*, 1999, **22**, 877.

85. V. V. Kelkar and K. M. Ng, *AIChE J.*, 1999, **45**.

86. M. Mazotti, A. Kruglov, B. Neri, D. Gelosa and M. Morbidelli, *Chem. Eng. Sci.*, 1996, **51**, 1827.

87. M. Maurer, U. Altenhöner, J. Strube and H. Schmidt-Traub, *J. Chromatogr. B*, 1997, **769**, 71.

88. M. Juza, M. Mazotti and M. Morbidelli, *Chromatografie*, 1998, **18**, 70.

89. A. Iwan, H. Stephenson, W. C. Ketchie and A. A. Lapkin, *Chem. Eng. J.*, 2009, **146**, 249.

90. K. K. Sirkar, P. V. Shanbhag and A. S. Kovvali, *Ind. Eng. Chem. Res.*, 1999, **38**, 3715.

91. S. L. Matson and J. A. Quinn, in *Membrane Handbook*, ed. W. S. Ho and K. K. Sirkar, Chapman & Hall, New York, 1992, p. 809.

92. D. W. Johnson, M. J. Semmens and J. Gulliver, *J. Membr. Sci.*, 1997, **128**, 67.

93. K. Brindle, T. Stephenson and M. J. Semmens, *J. Membr. Sci.*, 1998, **144**, 197.

94. G. T. Vladysavljevic, *Sep. Purif. Technol.*, 1999, **17**, 1.

95. A. A. Lapkin, B. Bozkaya and P. K. Plucinski, *Ind. Eng. Chem. Res.*, 2006, **45**, 2220.

96. P. Oxley, C. Brechtelsbauer, F. Ricard, N. Lewis and C. Ramshaw, *Ind. Eng. Chem. Res.*, 2000, **39**, 2175.

97. R. Jachuck, *Chem. Eng. Res. Des.*, 2002, **80**, 233.
98. C. Tsouris and J. V. Porcelli, *Chem. Eng. Prog.*, 2003, **99**, 50.
99. T. M. Squires and S. R. Quake, *Rev. Modern Phys.*, 2005, **77**, 977.
100. A. D. Stroock, S. K. W. Dertinger, A. Ajdari, I. Mezić, H. A. Stone and G. M. Whitesides, *Science*, 2002, **295**, 647.
101. V. Hessel, P. Angeli, A. Gavriilidis and H. Lwe, *Ind. Eng. Chem. Res.*, 2005, **44**, 9750.
102. S. Irandoust, B. Andersson, E. Bengtsson and M. Siverström, *Ind. Eng. Chem. Res.*, 1989, **28**, 1489.
103. S. Irandoust, S. Ertle and B. Andersson, *Can. J. Chem. Eng.*, 1992, **70**, 115.
104. G. Berčič and A. Pintar, *Chem. Eng. Sci.*, 1997, **52**, 3709.
105. G. Dummann, U. Quittmann, L. Groschel, D. W. Agar, O. Worz and K. Morgenschweis, *Catal. Today*, 2003, **79–80**, 433.
106. A. Günther, S. A. Khan, M. Thalmann, F. Trachsel and K. F. Jensen, *Lab Chip*, 2004, **4**, 278.
107. M. T. Kreutzer, P. Du, J. J. Heiszwolf, F. Kapteijn and J. A. Moulijn, *Chem. Eng. Sci.*, 2001, **56**, 6015.
108. T. A. Nijhuis, M. T. Kreutzer, A. C. J. Romijn, F. Kapteijn and J. A. Moulijn, *Chem. Eng. Sci.*, 2001, **56**, 823.
109. H. A. Smits, A. Stankiewicz, W. C. Glasz, T. H. A. Fogl and J. A. Moulijn, *Chem. Eng. Sci.*, 1996, **51**, 3019.
110. B. Ahmed, D. Barrow and T. Wirth, *Adv. Synth. Catal.*, 2006, **348**, 1043.
111. M. Sohrabi and A. M. Jamshidi, *J. Chem. Technol. Biotechnol.*, 1997, **69**, 415.
112. A. W. Kleingeld, L. Lorenzen and F. G. Botes, *Chem. Eng. Sci.*, 1999, **54**, 4991.
113. A. M. Dehkordi, *Ind. Eng. Chem. Res.*, 2001, **40**, 681.
114. F. G. Botes, L. Lorenzen and J. S. J. V. Deventer, *Chem. Eng. Commun.*, 1998, **170**, 217.
115. Y. Liu and R. O. Fox, *AIChE J.*, 2006, **52**, 731.
116. R. D. Chambers, D. Holling, R. C. H. Spink and G. Sandford, *Lab Chip*, 2001, **1**, 132.
117. N. d. Mas, A. Gunther, M. A. Schmidt and K. F. Jensen, *Ind. Eng. Chem. Res.*, 2009, **48**, 1428.
118. M. Zanfir, A. Gavriilidis, C. Wille and V. Hessel, *Ind. Eng. Chem. Res.*, 2005, **44**, 1742.
119. P. S. Kumar, J. A. Hogendoorn, P. H. M. Feron and G. F. Versteeg, *Chem. Eng. Sci.*, 2002, **57**, 1639.
120. A. A. Lapkin, S. R. Tennison and W. J. Thomas, *Chem. Eng. Sci.*, 2002, **57**, 2357.
121. M. W. Losey, R. J. Jackman, S. L. Firebaugh, M. A. Schmidt and K. F. Jensen, *J. Microelectromechanical Syst.*, 2002, **11**, 709.
122. E. V. Rebrov, A. Berenguer-Murcia, H. E. Skelton, B. F. G. Johnson, A. E. H. Wheatley and J. C. Schouten, *Lab Chip*, 2009, **9**, 485.

123. E. V. Rebrov, G. B. F. Seijger, H. P. A. Calis, M. H. J. M. d. Croon, C. M. v. d. Bleek and J. C. Schouten, *Appl. Catal., A: Gen.*, 2000, **206**, 125.

124. T. Conant, A. Karim and A. Datye, *Catal. Today*, 2007, **125**, 11.

125. S. K. Ajmera, M. W. Losey, K. F. Jensen and M. A. Schmidt, *AIChE J.*, 2001, **47**.

126. U. Kunz, H. Schönfeld, W. Solodenko, G. Jas and A. Kirschning, *Ind. Eng. Chem. Res.*, 2005, **44**, 8458.

127. B. Altava, M. I. Burguete, E. García-Verdugo, S. V. Luis and M. J. Vicent, *Green Chem.*, 2006, **8**, 717.

128. A. A. Lapkin, C. Savill-Jowitt, K. Edler and R. Brown, *Langmuir*, 2006, **22**, 7664.

129. S. Kataoka, A. Endo, A. Harada and T. Ohmori, *Mater. Lett.*, 2008, **62**, 723.

130. S. Polisski, B. Goller, A. Lapkin, S. Fairclough and D. Kovalev, *phys. stat. sol. (RRL)*, 2008, **2**, 132.

131. S. A. Alekseev, V. Lysenko, V. N. Zaitsev and D. Barbier, *J. Phys. Chem. C*, 2007, **111**, 15217.

132. E. M'ery, S. A. Alekseev, V. N. Zaitsev and D. Barbier, *Sens. Actuators. B*, 2007, **126**, 120.

133. J. Drott, K. Lindström, L. Rosengren and T. Laurell, *J. Micromech. Microeng.*, 1997, **7**, 14.

134. T. Saitoh, A. Sekino and M. Hiraide, *Anal. Chim. Acta*, 2005, **536**, 179.

135. Y. Voloshin, R. Halder and A. Lawal, *Catal. Today*, 2007, **125**, 40.

136. A. Pigamo, M. Besson, B. Blanc, P. Gallezot, A. Blackburn, O. Kozynchenko, S. Tennison, E. Crezee and F. Kapteijn, *Carbon*, 2002, **40**, 1267.

137. S. R. Tennison, *Appl. Catal., A: Gen.*, 1998, **173**, 289.

138. G. Dräger, C. Kiss, U. Kunzb and A. Kirschning, *Org. Biomol. Chem.*, 2007, **5**, 3657.

139. P. Lozano, E. García-Verdugo, R. Piamtongkam, N. Karbass, T. D. Diego, M. I. Burguete, S. V. Luis and J. L. Iborra, *Adv. Synth. Catal.*, 2007, **349**, 1077.

140. S. Kolaczkowski, P. Plucinski and A. A. Lapkin, *Récents Progrès en Génie des Procédés,* SFGP, Paris, France, 2007.

141. G. G. Haselden, US Pat., 3 528 783, 1970.

142. O. Goerke, P. Pfeifer and K. Schubert, *Appl. Catal., A: Chem.*, 2004, **263**, 11.

143. H. Lu, M. A. Schmidt and K. F. Jensen, *Lab Chip*, 2001, **1**, 22.

144. S. Meyer, D. Tietze, S. Raub, B. Schäfer and G. Kreisel, *J. Photochem. Photobio. A: Chem.*, 2007, **186**, 248.

145. J. C. McDonald and G. M. Whitesides, *Acc. Chem. Res.*, 2002, **35**, 491.

146. Z. T. Cygan, J. T. Cabral, K. L. Beers and E. J. Amis, *Langmuir*, 2005, **21**, 3629.

147. R. Knitter and M. A. Liauw, *Lab Chip*, 2004, **4**, 378.

148. R. Knitter, D. Göhring, P. Risthaus and J. Haußelt, *Microsystem Technol.*, 2001, **7**, 85.
149. S. L. Poe, M. A. Cummings, M. P. Haaf and D. T. McQuade, *Angew. Chem. Int. Ed.*, 2006, **45**, 1544.
150. J. M. Commenge, L. Falk, J. P. Corriou and M. Matlosz, *AIChE J.*, 2002, **48**, 345.
151. E. R. Delsman, A. Pierik, M. H. J. M. D. Croon, G. J. Kramer and J. C. Schouten, *Chem. Eng. Res. Des.*, 2004, **82**, 267.
152. E. V. Rebrov, I. Z. Ismagilov, R. P. Ekatpure, M. H. J. M. d. Croon and J. C. Schouten, *AIChE J.*, 2007, **53**, 28.
153. E. V. Rebrov, R. P. Ekatpure, M. H. J. M. d. Croon and J. C. Schouten, *J. Micromech. Microeng.*, 2007, **17**, 633.
154. T. Bayer, J. Jenck and M. Matlosz, *Chem. Eng. Technol.*, 2005, **28**, 431.
155. H. Krummradt, U. Kopp and J. Stoldt, *Proceedings of the 3rd International Conference on Microreaction Technology*, Berlin, 2000.
156. Y. Matsushita, S. Kumada, K. Wakabayashi, K. Sakeda and T. Ichimura, *Chem. Lett.*, 2006, **35**, 410.
157. R. Gorges, S. Meyer and G. Kreisel, *J. Photochem. Photobiol. A: Chem.*, 2004, **167**, 95.
158. G. Takei, T. Kitamori and H.-B. Kim, *Catal. Comm.*, 2005, **6**, 357.
159. J. Berna, D. A. Leigh, M. Lubomska, S. M. Mendoza, E. M. Perez, P. Rudolf, G. Teobaldi and F. Zerbetto, *Nat. Mater.*, 2005, **4**, 704.
160. X. Fan, H. Chen, Y. Ding, P. K. Plucinski and A. A. Lapkin, *Green Chem.*, 2008, **10**, 670.

Flow Processes Using Polymer-supported Reagents, Scavengers and Catalysts

EDUARDO GARCÍA-VERDUGO AND
SANTIAGO V. LUIS

Department of Inorganic and Organic Chemistry, University Jaume I, Avda.
Sos Baynat s/n, E-12071, Castellón, Spain

2.1   Introduction

The attachment of functional organic moieties onto solid, insoluble matrices in
order to develop solid-supported reagents, scavengers and catalysts has been a
very useful approach for many years for the development of facilitated chemical syntheses and, herein, that of more environmentally benign chemical
processes.[1–3] A number of distinct advantages are associated with these supported systems:

  (i) easy separation from the reaction medium through simple filtration
      procedures;
 (ii) possibility of regeneration of supported reagents and scavengers;
(iii) possibility of recycling of supported catalysts;
 (iv) reduction of toxicological concerns associated with the use of toxic
      reagents and catalysts;
  (v) facilitation of work-up providing, for instance, odourless materials.

RSC Green Chemistry No. 5
Chemical Reactions and Processes under Flow Conditions
Edited by S.V. Luis and E. Garcia-Verdugo
© The Royal Society of Chemistry 2010
Published by the Royal Society of Chemistry, www.rsc.org

An additional advantage, in particular in light of the original work on solid-phase synthesis[4,5] and with the perspective of combinatorial chemistry,[6,7] is the suitability of those materials for automatisation.[8] More recently, the potential of solid-supported reagents, scavengers and catalysts for developing flow processes has been realised as one of their key properties.[9]

The matrices used as the supports can be classified, according to their nature, into organic and inorganic. In this chapter, we concentrate solely on the use of organic, polymeric matrices and the use of polymer-supported reagents, scavengers and catalysts for developing chemical processes under flow conditions.

Many different kinds of crosslinked, insoluble polymers have been studied as supports for the immobilisation of reactive organic functionalities.[10–12] Nevertheless, most of the work has concentrated on the use of two relatively simple polymeric networks: crosslinked polystyrene (PS) and polyacrylic (PA) derivatives (Figure 2.1). Polymeric materials derived from polystyrene crosslinked with variable amounts of divinylbenzene (PS-DVB) are by far the most common supports for the development of polymer-supported reagents, scavengers and catalysts.

A number of advantages have been ascribed to the use of organic instead of inorganic supports including:[13]

- the easier functionalisation of organic matrices containing well-defined organic functionalities (*i.e.* the aromatic rings in PS-DVB);
- the higher loadings attainable;
- the chemical stability to certain conditions such as the use of strong acids and bases;
- the possibility of controlling the nature and properties of the polymeric matrix itself through the appropriate modification of the original polymerisation conditions.

Of course, several disadvantages can also be reported—in particular the much lower thermal stability associated with organic matrices. Thus, the selection of an organic or inorganic support needs to be made in the context of the specific application to be carried out.

Two main categories of polymer-supported species can be considered based on the use of soluble or insoluble matrices. Soluble polymers can be very different in nature and include the use of lineal not-crosslinked polymers and oligomers such as polyethyleneglycol (PEG) and polyethylene (PE),[14,15] or more complex structures such as dendrimers.[16] The use of these soluble species for flow applications is associated, in principle, with membrane reactors in which the polymer-bound species are retained on one of the sides of the membrane while the products and reactants flow through. This field has been reviewed in detail[17] and we will not concentrate on it here.

For insoluble polymers, the morphology of the support is always an important parameter for defining their applications.[18] This is particularly

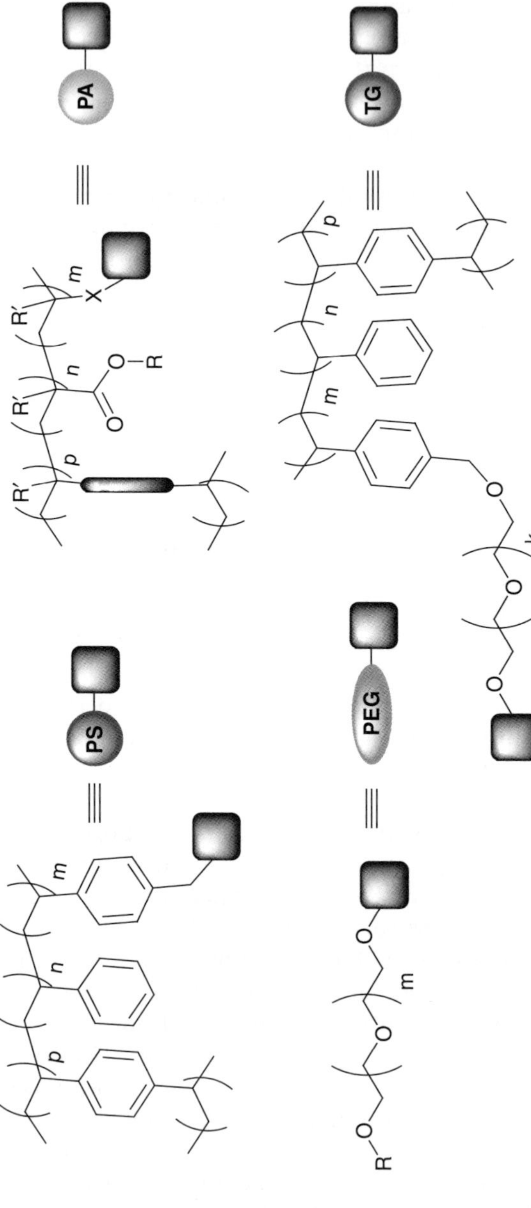

Figure 2.1 General chemical structures for the most common polymeric supports: polystyrene-divinylbenzene (PS), polyacrylate (PA), oligoethylene glycol (PEG) and Tentagel™ (TG).

critical for developing flow systems. In this regard, the functional resins can be classified into three main categories:[13,19]

(i) insoluble beads of a gel-type;
(ii) macroporous beads;
(iii) macroporous monoliths.

Microporous, gel-type beads are prepared through suspension polymerisation and have a very low degree of crosslinking. Thus, in the case of polystyrene-divinylbenzene copolymers, the crosslinking agent (DVB) usually makes up 1–2% of the monomeric mixture. Macroporous beads are also obtained by suspension polymerisation, but have a higher content of the crosslinker and the polymerisation is carried out in the presence of a porogenic agent. The preparation of macroporous monoliths requires the use of a mould in which the monomeric mixture, along with the porogenic agents, is poured. The polymerisation takes place inside the mould, usually—but not always—by thermal radical polymerisation. This requires the use of a high content of crosslinker and allows the polymer to be obtained as a single piece (monolith).[20]

Each morphology presents advantages and disadvantages from the perspective of their use under flow conditions. Accessibility to the active sites is one of the main concerns when developing polymer-supported systems.[21] By definition, gel-type resins do not possess any permanent porosity and thus their swelling in an appropriate solvent is required for any reaction to occur in the interior of the beads. The selection of the solvent depends on the chemical nature of both the polymeric backbone and the functional groups being introduced. Thus solvents such as toluene or dichloromethane are very appropriate for PS-DVB matrices, while more polar solvents are compatible with acrylic-derived resins. The need to produce PS-DVB resins compatible with alcohols or even water has led to the development of new families of polymers in which the polymeric backbone contains additional functional groups—in particular oligoethylene glycol moieties—either as crosslinkers or as linkers between the matrix and the functional group.[22]

The use of long spacers between the matrix and the functional group has been a general strategy to improve the accessibility of the active site in this kind of resin.[23] However, from the point of view of flow applications, it is important to bear in mind that the swelling process is accompanied by an increase in the size of the beads which can lead to a significant change in the overall volume of the resin being used.[24] This needs to be considered when designing the reactor configuration to avoid overpressures, mechanical damage of the resin, *etc.*

The importance of swelling is illustrated in Figure 2.2, where the behaviour of three different resins (1% crosslinked gel-type resin, 50% crosslinked macroporous beads and 60% crosslinked macroporous monolith) after contact with dichloromethane can be visualised. Under some circumstances, the increase in volume produced by swelling for the gel-type resins can reach values of up to 200%.

Figure 2.2 PS-DVB polymers with different morphologies. From left to right: (a) gel-type resin (1% crosslinked), (b) macroporous resin (5% crosslinked), (c) monolith.

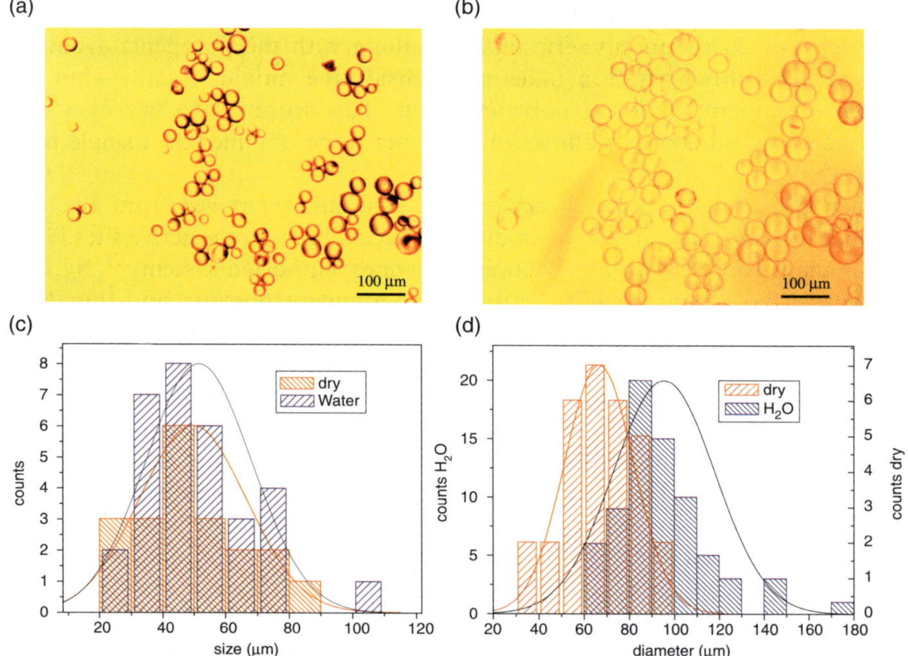

Figure 2.3 Swelling in water of (a) a gel-type Merrifield polymer resin and (b) a similar resin with the chlorine groups substituted by methylimidazolium groups. (c) and (d) represent the corresponding size distribution histograms for resins depicted in (a) and (b).

The importance of the functional groups present in the matrix for the swelling properties of the resulting resins is clearly shown in the example in Figure 2.3. In this case, the swelling properties in water of a 1% crosslinked PS-DVB Merrifield polymer resin (chloromethylated PS) are compared with those of the same resin after transforming the chloromethyl groups into imidazolium functionalities.[25,26] In this case, the introduction of a highly polar functional group allows good swelling in water, providing an increase in size of *ca.* 73%, while the original polymer is completely hydrophobic and does not swell in water.

In the case of macroporous resins with a high degree of crosslinking, swelling does not usually produce a change in volume. The presence of a permanent porous structure on the beads or on the monoliths facilitates the flow-through of solvents and the accessibility of the surfaces of the porous structure. Nevertheless, the polymer regions not located at these surfaces can be much less accessible to reagents and substrates, as the higher crosslinking precludes proper swelling of those regions.

For macroporous polymers, the role of the porogenic agents is critical as the chemical nature and the amount used of these agents, which are not incorporated into the final structure, determine the porosity of the resulting polymer.[13,27] Although macroporous resins have a higher mechanical stability than gel-type resins, they are also somewhat brittle and this also needs to be taken into account when considering their potential applications.

From the point of view of reactor design, it is important to be aware that packing of macroporous beads can be relatively inefficient, producing large dead volumes. On the other hand, a more efficient packing is accompanied by a build up of pressure, which can be an inconvenient from the point of view of the engineering of the process and for the mechanical stability of the resin. These problems can be avoided with the use of a monolith as a single piece adjusted to the shape and volume of the reactor and having a porosity designed to allow proper flow of solvents and reagents at reasonable pressures. Differences in packing using monoliths or bead resins are illustrated in Figure 2.4, which shows the change in the flow pattern attained as a consequence of the presence of preferential channels in the case of packed beads columns (b).[28]

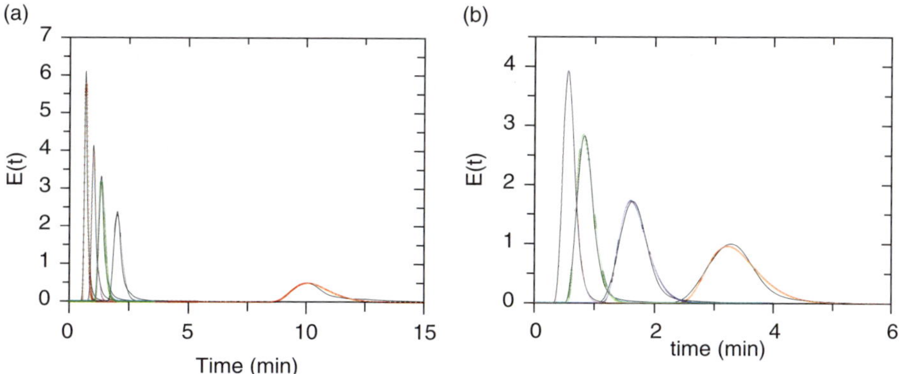

Figure 2.4   Comparison of flow patterns observed for a monolithic column (a) and a column packed with bead-type polymers and showing the presence of preferential channels (b).

2.2   Flow Processes with Use of Bead-type Resins

The original design of Merrifield's peptide synthesis contained the potential for working under flow conditions in an automatic system.[29] For many years, however, the strict application of flow conditions in work with functionalised polymers was hampered by the misconception of reactions taking place in polymers requiring very long reaction times. Thus, typical protocols for transformation of functional moieties in resins involved reactions times of 48 hours.[30] Processes having very slow kinetics are always difficult to be adapted to flow conditions; nevertheless, the requirement of such long reaction times has been demonstrated to be unnecessary in many cases.[26] Although diffusion problems can in some instances reduce the rate of a given reaction compared to that in solution, appropriate selection of the matrix, the solvent and the reaction conditions very often allows those limitations to be overcome. Indeed, appropriate design of the system can allow the development of supported reagents and catalysts that are more active than the corresponding homogeneous ones.[31]

The use of long reaction times with polymer-supported reagents, scavengers and catalysts is often based on the absence of efforts to optimise the reaction conditions. Originally, this was due to the limited number of analytical tools able to monitor the progress of a reaction on a solid support. Thus, for example, the quantitative transformation of chloromethyl groups on a Merrifield resin (chloromethylated PS-DVB) into imidazolium groups (Scheme 2.1) has been reported to require 24–72 hours.[32] However, simple optimisation of the process based on its monitoring using different techniques allowed the reaction time to be reduced to a few minutes.[26] The same is clearly true for reactions taking place in solution with the help of a supported reagent, scavenger or catalyst; thus, in many cases, there are no rate limitations when transferring processes using functionalised resins from batch to flow.

2.2.1   Use of Gel-type Beads

Following the seminal work by Merrifield,[29] bead gel-type resins were the first ones to be used in the area of facilitated chemical synthesis. Thus, they are likely to be the first ones to be assayed for the development of a flow process; but, for this to be possible, proper design of reactor configuration is essential.[33]

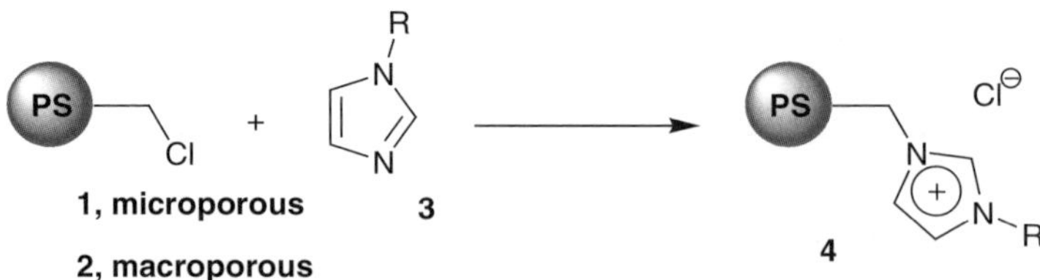

Scheme 2.1   Synthesis of imidazolium-containing polymers (**4**) from Merrifield resins (**1** and **2**) and imidazole (**3**).

This is vital if we consider the importance that changes in the size of the particles that take place on swelling can have in defining the flow patterns of the resulting reactors.

The packing of a column (a simple packed-bed reactor) with functionalised gel-type beads is the first approach toward the design of flow processes. In this regard and taking into consideration our earlier discussion, it is essential that the packing of the column takes account of the estimated change in volume that will take place following swelling of the polymer with the solvent to be used for the process. As in any other packed column, careful control is required to avoid the formation of cracks or floating of beads, which could lead to an inhibition of flow through the beads and contact of the substrates and reagents with the active sites at the internal surfaces of the resin.

Three classical reactor designs (types I–III) have been described using this approach; these are described in detail in ref. 33 and schematically represented in Figure 2.5. First examples of the application of functionalised resins to flow

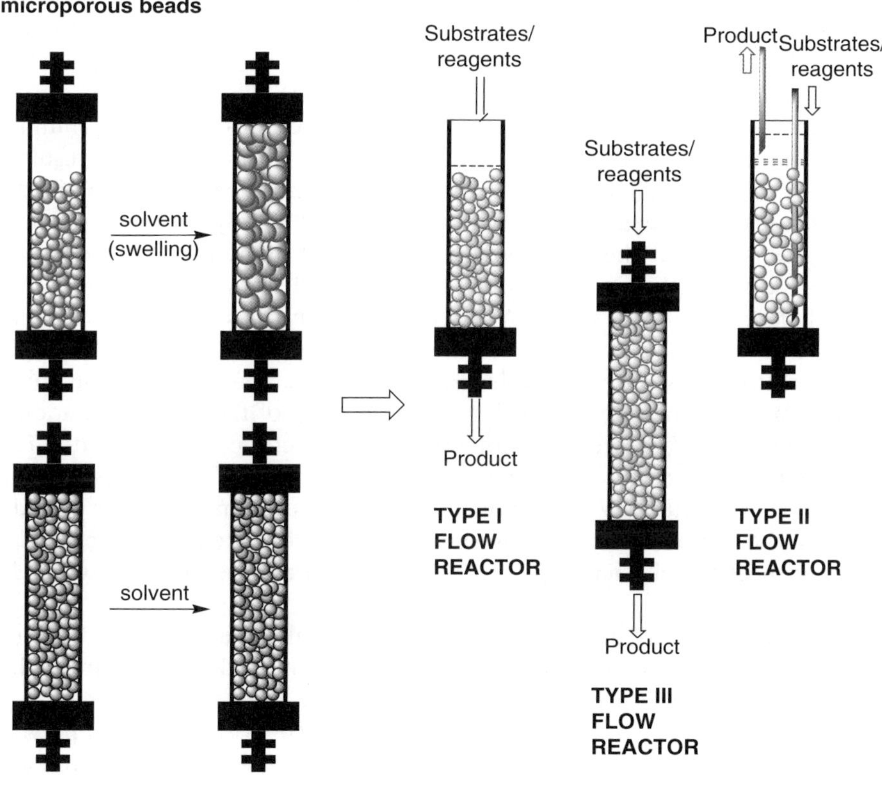

Figure 2.5   Different reactor configurations used with polymer-supported scavengers, reagents or catalysts.

processes used the simplest design based on gravity flow columns (type I). In this case, the solution(s) containing the reactants (and reagents) are fed to the top of the column and the flow through the column is driven by gravity. The assembly depicted as a type II reactor represents an equivalent to fluidised bed reactors. The solution(s) containing the reactants (and reagents) are fed to the bottom of the column and flow upwards; here the volume of the reactor must be higher than the volume of the swollen polymer to account for the fluidisation. The solution containing the product(s) is collected at the top of the column by overflow or with the use of an additional device.[33] The use of a pump to control the flow through the column provides access to more elaborate systems (type III reactors). Those systems provide a more precise control of the flow, temperature and other reaction parameters than type I and II arrangements.

The need for proper adjustment of reactor design has hampered the use of gel-type resins for efficient flow processes on many occasions and examples have been reported in which a change from microporous to macroporous has been necessary to prevent column blockage due to polymer swelling.[34] Accordingly, only a limited number of examples can be found in the literature.

Itsuno *et al.* developed different amino acid derived PS-supported catalysts such as **5** and **6** (Figure 2.6).[35] The borane complex of the supported amino alcohol **5** was used for borane reduction of ketones in tetrahydrofuran (THF) using a fluidised bed reactor configuration (type II). The mixture of reagent (borane) and substrate (ketone) was pumped to the bottom of the column at flow rates of 0.4–$0.8\,\mathrm{mL\,min^{-1}}$ (residence time 30–60 min) and the resulting solution was collected by overflow. The corresponding alcohol was obtained in excellent yields with enantioselectivities of up to 93%. Polymer **6** was prepared by polymerisation of the corresponding functional monomer and a relatively high degree of crosslinking (10%), but using a flexible crosslinker containing an oligoethylene glycol fragment.[36] A gravity flow set-up was used and the borane complex of the polymer-bound *N*-sulfonyl amino acid catalysed the Diels–Alder reaction between methacrolein and cyclopentadiene in dichloromethane (DCM). With a flow rate of $10\,\mathrm{mL\,h^{-1}}$, up to 18.8 g of the mixture of adducts were obtained (95% yield, endothermic : exothermic = 93:7) with 71% enantiometric excess (ee). This illustrates the possibility of using these systems to increase production by scaling-out (running the reaction for longer periods). For this, the use of flow conditions under the appropriate set-up is critical to allow the mechanical, chemical and fluid dynamic properties to be preserved for long-term use.

The enantioselective addition of diethyl zinc to aldehydes catalysed by supported chiral amino alcohols is probably the most studied reaction under flow conditions. Examples using gel-type resins include the use of polymers **7–10** (Figure 2.6). The first study by Fréchet, Itsuno and co-workers derived polymer **7** from resin **5**.[37] Polymers with variable amounts of crosslinking (up to 20%) and functionalisation degrees were prepared and assayed. A fixed bed configuration (type I reactor) was used and the reaction at $0\,^{\circ}\mathrm{C}$ afforded, after a single pass, the alcohol **15** (Ar = 4-Cl-phenyl) in 90% yield and 94% ee (*S* configuration for the major enantiomer).

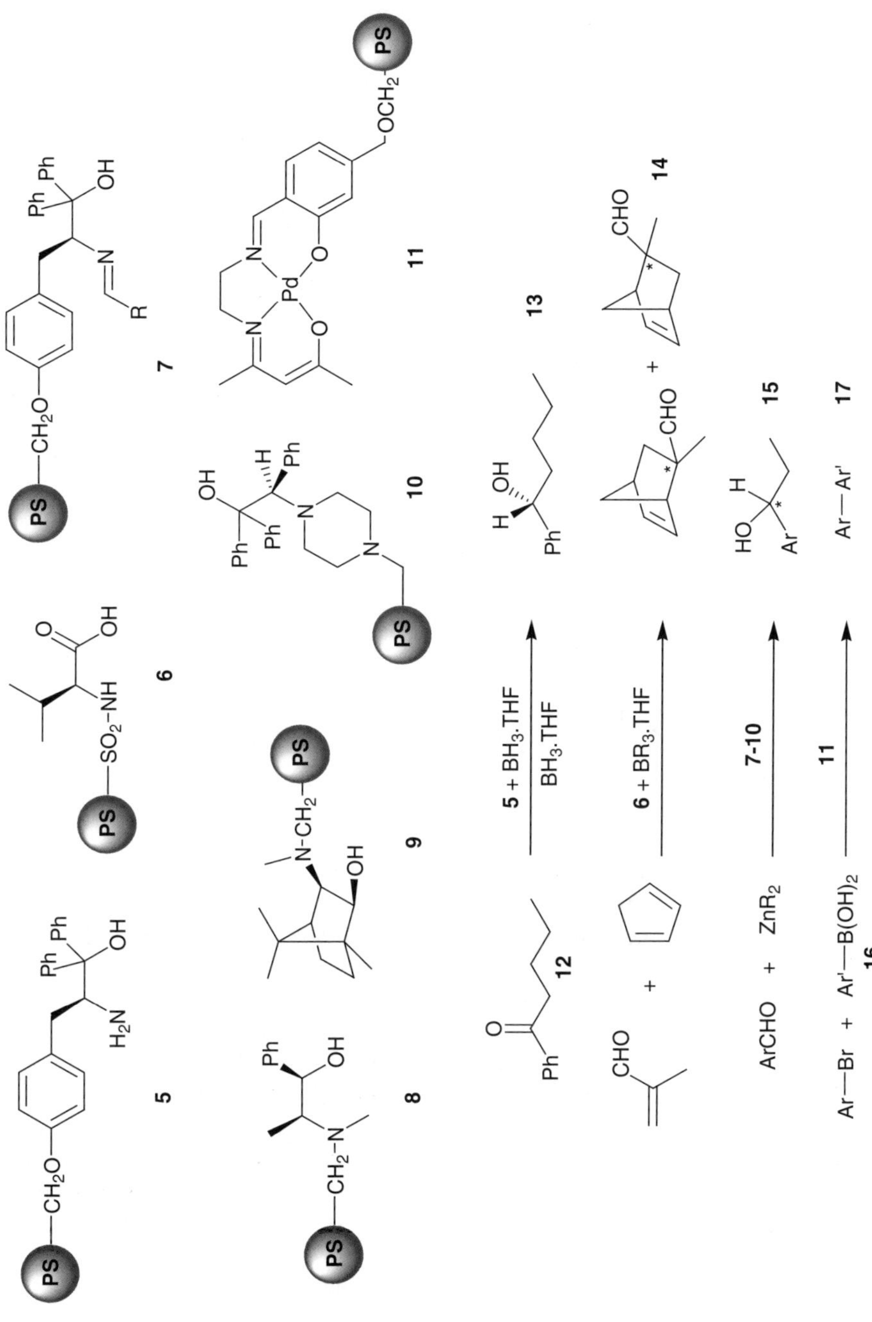

Figure 2.6 Some microporous (gel-type) polymer-supported catalysts for continuous flow C–C bond formation reactions.

This reaction was studied in detail by Hodge *et al.* using different supported amino alcohols under different conditions.[38] Using fluid bed flow (type II) systems, they found that the outcome of the reaction (yields and selectivities) depended significantly on the structure of the amino alcohol and on the concentrations and flow rates of the soluble reactants. With polymer **8**, at 20 °C and flow rates of 10 mL h^{-1}, yields of 98% (with 98% ee, *R* as the major isomer) were obtained for benzaldehyde.[38] It is worth mentioning that for these experiments, resins with a very low level of crosslinking were used (0.2%). The fragility of these polymers makes efficient use difficult under batch conditions, but they could be repeatedly used in the flow system. With polymer **9**, using a flow rate of 6 mL h^{-1}, the flow system was used continuously for up to 275 hours, affording a 95% yield (97% ee, *S* as the major isomer) for benzaldehyde.[38] This is remarkable taking into account the high sensitivity to air and humidity of the reagent. Nevertheless, it has to be mentioned that a decrease in the enantioselectivity was observed for successive runs and this was ascribed to the partial epimerisation of the supported ligand.[38]

More recently, Pericàs *et al.* studied the addition of diethyl zinc to benzaldehydes under flow conditions with the use of polymer **10**.[39] For this purpose, a low-pressure chromatographic column was filled completely with the swollen resin and a type III configuration was set up. Two high-pressure pumps provided the flow of the substrate and reagent through the column. The high reactivity of this catalyst allowed the reaction to be carried out (97% yield, 93% ee, *S* as the major enantiomer) with very short residence times (2.8 min).

The use of the polymer-bound non-chiral Pd-salen analogue **11** (Figure 2.6) has been studied under flow conditions with a type III configuration for the Suzuki–Miyaura reaction between different aryl bromides and phenyl boronic acid.[40] The authors found a 20 times enhancement in reaction rates compared with the batch reactor. This has been attributed, as in other related examples, to the fact that continuous flow reactors provide an actual catalyst–substrate ratio much higher than under traditional batch conditions and one that can be regulated by adjusting the flow. The use of stop-flow techniques was also considered to increase the conversion from 60% to 86%. This might be a useful approach for reactions displaying very slow kinetics.

Basic catalysis can be carried out efficiently with the use of polymer-bound bases such as the commercially available resin PS-BEMP **18**.[41] (Figure 2.7). This resin has been reported to catalyse efficiently the Michael addition of nitroethane to *E*-benzylideneacetone (**20**). A gravity flow system set-up was used under solventless conditions. However, recirculation for four hours of the mixture of reagents and products coming out of the column was required to achieve full conversion of 50 mmol of **20** using a 5% molar loading of the catalyst. The catalyst was found to have limited stability, as regeneration was needed after the third cycle.

The attachment of cinchona alkaloids to Merrifield resins *via* sulfide linkages allowed Hodge and Cazé and co-workers to prepare a series of chiral organocatalysts such as **19** that were assayed for the synthesis of adduct **22** through a Michael addition.[42] The authors set up a type II assembly for flow studies and

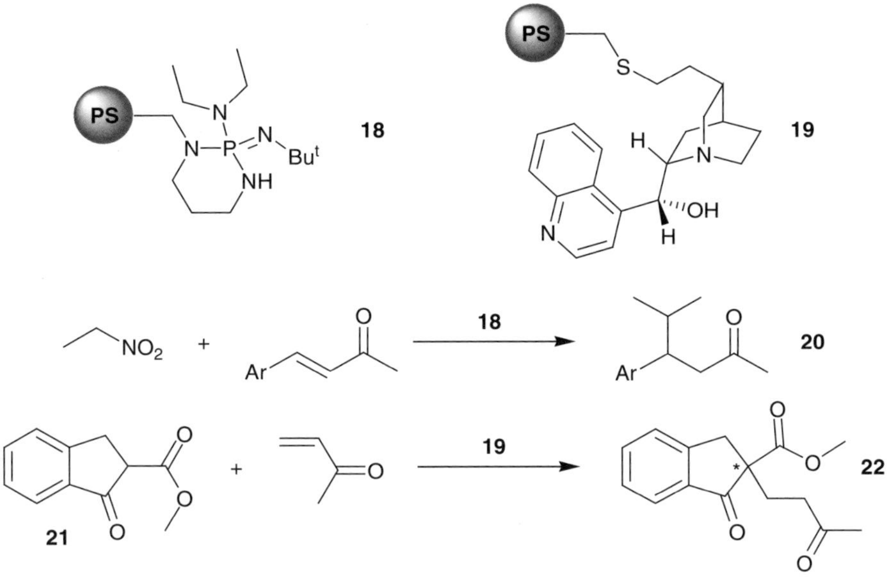

Figure 2.7   Basic microporous polymer-supported catalysis.

found that long residence times (6 h, flow rate $5\,mL\,h^{-1}$) were required to achieve yields of 96% with enantioselectivities of 50–52%.

2.2.2   Use of Macroporous Beads

The use of macroporous beads instead of microporous ones avoids some of the practical limitations considered above as the swelling does not produce significant changes in volume. Thus, it is not surprising that many of the earlier reports on the use of functionalised polymeric beads under flow conditions describe the use of macroporous systems, including the applications based on macroporous ion exchange resins discussed below. On the other hand, these characteristics favour the use of type III reactor assemblies with a packed bed arrangement and flow-through based on the use of different pumping devices.

Many different kinds of supported reagents and scavengers have been reported, mainly using batch conditions.[43] However, additional issues need to be taken into account for their application under flow conditions. In particular, it must be borne in mind that the supported reagents and scavengers have a finite capacity that gradually reduces with use; after a certain time they need to be regenerated or replaced.[9,33]

To this category belong the oxidation reactions reported by Hodge and Harrison in the 1970s and 1980s. Thus, the oxidation of the sulfide group in a family of penicillins was carried out under flow conditions using a PS-supported peroxyacid in a fixed bed column heated at 40 °C.[44] Using a residence time of 30 minutes, a 91% yield of the sulfoxide **26** was obtained (Figure 2.8).

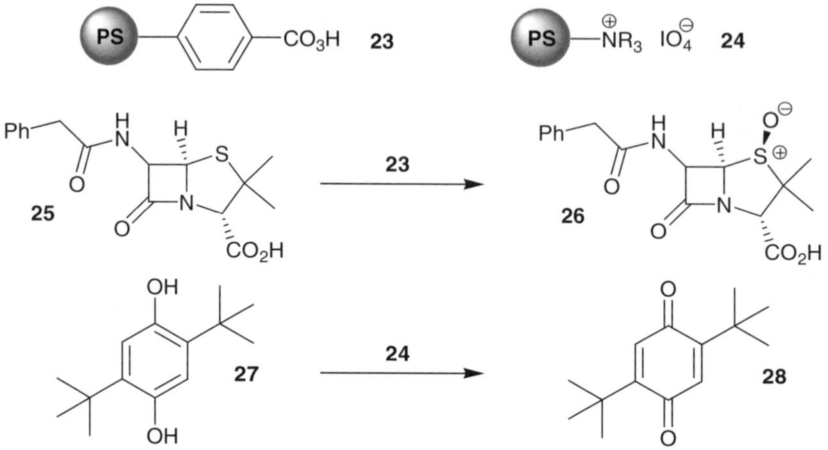

PS—⟨C$_6$H$_4$⟩—CO$_3$H   **23**   PS—$\overset{\oplus}{\mathrm{N}}R_3$ IO$_4^{\ominus}$ **24**

Figure 2.8 Stoichiometric oxidation reaction using polymer-supported reagents under flow conditions.

This compares well with the 85% yield for a normal batch experiment for 2 h at 20 °C with 1.5 equivalents of resin. After reaction, the polymer with a functional degree of $1.9\,\mathrm{mmol\,g^{-1}}$ could be regenerated efficiently using hydrogen peroxide and methanesulfonic acid (12 h, 20 °C); the regenerated reagent recovered its original activity fully. The second example also used a fixed bed arrangement (type I reactor).[45] In this case, resin **24** containing periodate fragments was prepared from a commercial anion exchange macro-porous resin and was efficient for the oxidation of di-*tert*-butylquinol **27** at 20 °C (97% yield) (Figure 2.8).

Starting from a commercially available methacrylate based Amberlite Oxi-rane resin, McQuade *et al.* developed the organocatalytic resin **29** (Figure 2.9) and packed it on a fluoroelastomeric tubing (1.6 mm inner diameter).[34] The use of a syringe pump ($50\,\mu\mathrm{L\,min^{-1}}$) afforded a type III assembly. The Knoevenagel preparation of compound **34** was carried out at 60 °C with 93% yield and a residence time of 280 s. The same authors also prepared the polymer-bound DMAP analogue **30**. This resin was used under flow conditions for the acyla-tion of phenyl ethanol (**35**). The reaction was carried out at a flow rate of $0.5\,\mathrm{mL\,min^{-1}}$ (residence time of 48 s) to afford full conversion of the starting material. In both cases, the productivity of the flow reactions was higher than that obtained for batch processes under similar conditions. This is a feature that is common to most flow processes studied up to now.

A similar approach has been reported to immobilise Tempo subunits for the biphasic oxidation of alcohols to aldehydes and ketones. It is important to note that, in some instances, for batch systems the resins showed signs of structural damage.[34] The copper complex derived from the macroporous PS-supported bisoxazoline **31** (Figure 2.9) was used by Salvadori *et al.* as an enantioselective catalyst for the glyoxate–ene reaction using a type III arrangement.[46]

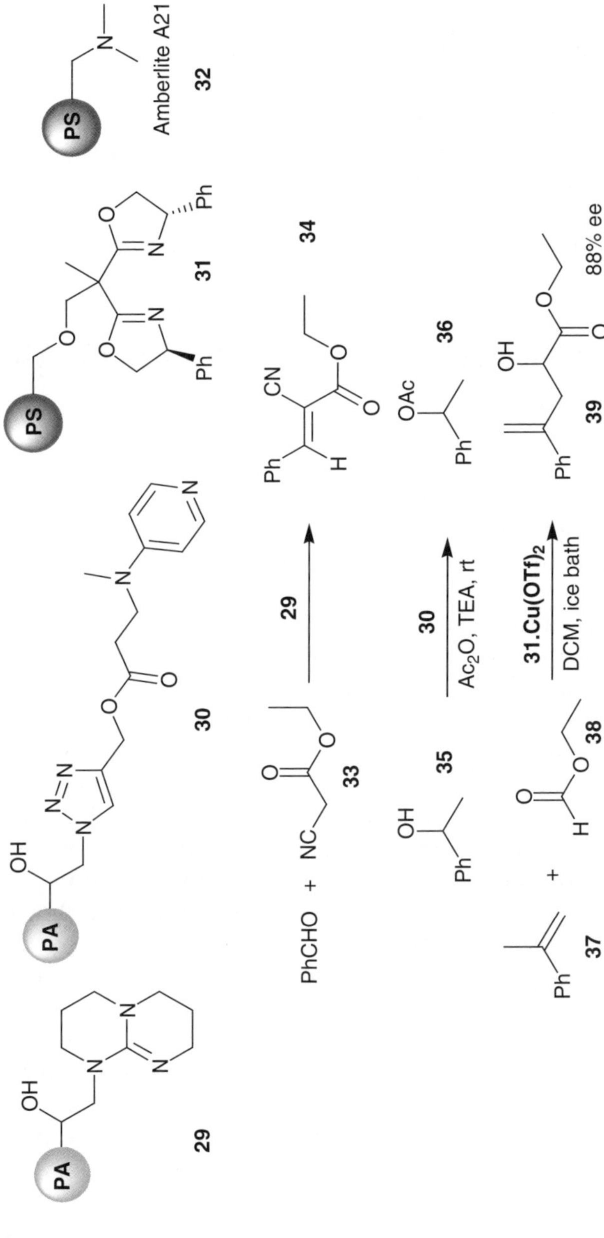

Figure 2.9 Bead-type macroporous polymer-supported catalysts.

The adduct **39** was obtained in DCM/toluene at the temperature provided by an ice bath, in 78% yield with 88% ee. Washing of the column with DCM was required between successive runs. A chiral salen ligand attached to macroporous divinylbenzene beads modified with a hydrophilic coating was used by Moberg *et al.* to prepare a Ti catalyst for the enantioselective cyanation of benzaldehyde under flow conditions (0.8 µL min^{-1}, 92% conversion, 72% ee).[47]

Simple, commercially available, basic anion exchange resins such as Amberlite A21 (**32**) have been also studied, for instance, for the Michael reaction leading to **22**.[42] Using a type II arrangement and flow rates of 5–21 mL h^{-1} (residence times 40–42 h), yields obtained ranged from 14 to 99%; the working temperature was a key factor for the efficiency of the process.[42]

2.3   Flow Processes with Use of Polymeric Monoliths

2.3.1   General Remarks

A polymeric monolith can be described as a continuous polymeric phase containing a permanent porous (macroporous) structure.[48] In the most general approach, developed by Fréchet,[49] a mixture of the corresponding monomers, along with a mixture of porogenic co-solvents, is poured into a mould where polymerisation is carried out in the absence of stirring. A high degree of crosslinking is used to ensure that the shape and morphology of the monolith is maintained after the synthetic protocol. A number of factors (*e.g.* temperature and concentration of the initiator) affect the characteristics of the permanent porous network created after removal of the porogenic mixture. However, the concentration and chemical nature of this porogenic mixture are the key parameters determining the porosity present in the final monolith.[27,49,50] In general, a high content of porogens (40–60% of the final mixture) is used and, typically, a series of experiments is required to determine the optimum conditions and solvents. When working under flow conditions, mean diameter pores of about 1000 nm are preferred to ensure a turbulent flow regime and to avoid excessive back pressures.

Although initial studies focused on radical-initiated preparation of polystyrene matrices, other monomeric structures have been also described. Of particular interest are the monolithic materials prepared by radical copolymerisation of glycidyl methacrylate and other acrylates, which have been used for application in chromatography, solid-phase synthesis and biocatalysis.[20,49]

In the present context, several distinct advantages can be associated with the use of monolithic polymers:[9,19,20,27]

(i) Diffusion problems are reduced, in particular in the presence of non-swelling solvents.

(ii) Accessibility to the active sites is improved due to the diffusion factors and the turbulent flow regime achieved.

(iii) Higher mechanical stability allows the monoliths to maintain their shapes and morphologies under a broad range of condition. This is essential for technical application and long-term stability.

(iv) A large variety of structures can be prepared by the polymerisation of the appropriate monomeric mixtures involving functional and structural monomers.

(v) The use of different moulds allows the preparation of monoliths with different shapes according to the technical needs for the particular flow process.

With regard to the last point, rods and discs are the most common shapes considered for polymeric monoliths to be used in flow processes.[19–20,27,51] In the case of the rods, direct polymerisation inside glass or steel columns like the ones used for chromatographic applications simplifies the assembly of the experimental set-up required for flow techniques.

Different alternative approaches have been used for the preparation of functionalised monolithic polymers. Thus, Kirschning lead the development of the so-called PASSflow reactors (Polymer Assisted Solution Phase Synthesis in the Flow through mode).[52] In this case, the polymerisation is carried out inside the microchannels of an inert inorganic support such as highly porous glass rod; different assemblies have been prepared based on these porous materials containing a monolithic polymeric phase. A second approach for the production of highly permeable polymeric monoliths was reported through the polymerisation of a continuous phase of a High Internal Phase Emulsion (HIPE). This produces a polymeric material named PolyHIPE, which has an open cellular structure and surface areas of *ca.* $700 \, \text{m}^2 \, \text{g}^{-1}$.[53] On the other hand, Buchmeiser *et al.* developed a ring-opening-methatesis (ROMP) strategy for the preparation of polymeric monolithic materials.[54] This living polymerisation approach allows a polymer functionalised with a second-generation Grubbs catalyst to be obtained directly.[55] A final alternative methodology was reported by Hillmyer *et al.*, with the preparation of monolithic materials containing nanoscopic polylactide (PLA) embedded in polystyrene.[56] Removal of PLA affords a monolith with well-defined nanochannels. The use of those materials is limited, however, because of their low mechanical and chemical stability.

2.3.2 Monolithic Reagents and Scavengers

The use of monolithic materials under flow conditions is always associated with a type III reactor design, though different experimental approaches can be used. Polystyrene polymeric discs containing acylating moieties according to structures **40** and **41** (Figure 2.10) were used for the acetylation of amines under mild conditions.[57] For the flow set-up the discs were encased in a polyethylene ring and inserted into a custom-made holder.[58] Quantitative acylation could be obtained with the use of a three-fold excess of acylating moieties (flow rate: $3 \, \text{mL} \, \text{h}^{-1}$). Different disks can be stacked on the top of each other to increase

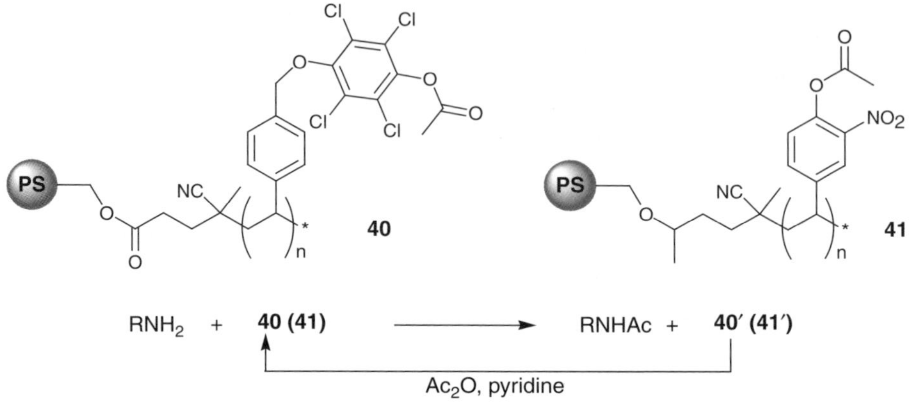

Figure 2.10 Polymer-supported acylating agents.

the overall capacity of the system or even to prepare multitask assemblies. In this regard, the isolation of the pure amide is simplified by the combination of acylating disks with scavenging disks containing functionalities able, when conversion is not complete, to remove the unreacted amines.

Polymers **40** and **41** are prepared following a three-step protocol. First, the PS monolith with the appropriate shape and morphology is prepared and functionalised to anchor a free radical azo initiator, which allows the initiation of a graft polymerisation to introduce linear polymeric chains on the surface of the monoliths. Finally, further modification of these linear polymeric chains is used to introduce the final functional groups.

In the case of PASSflow reactors, the preparation of supported quaternary ammonium salts with different counteranions allowed the study of a variety of reactions according to the nature of the specific anion. Thus, the use of nucleophilic anions such as phenoxide provided access to substitution reactions for the synthesis of alkyl aryl ethers **44** (Figure 2.11), while redox processes were carried out with the use of borohydride (reduction) or bisacetoxy-bromate(I) in the presence of catalytic TEMPO (oxidation). Finally, the hydroxide form can be used for the Horner–Emmons olefin synthesis and, in combination with [Pd(PPh₃)₄], as the base for the Suzuki reaction.[52,59] To avoid the limitations associated with slow reaction kinetics, the systems were operated with recirculation of the solution. All the former stoichiometric transformations were cleanly carried out in quantitative yields in most cases, the product being isolated merely by removal of the solvent. Only for the Suzuki transformation were yields lower.

Monolithic materials polymerised inside chromatographic glass columns containing similar triethyl ammonium functionalities have been developed by Ley and co-workers.[60] The azide form of this polymer, prepared by pumping an aqueous sodium azide solution through the column, has been applied for the preparation of acyl azides that subsequently experienced the Curtius

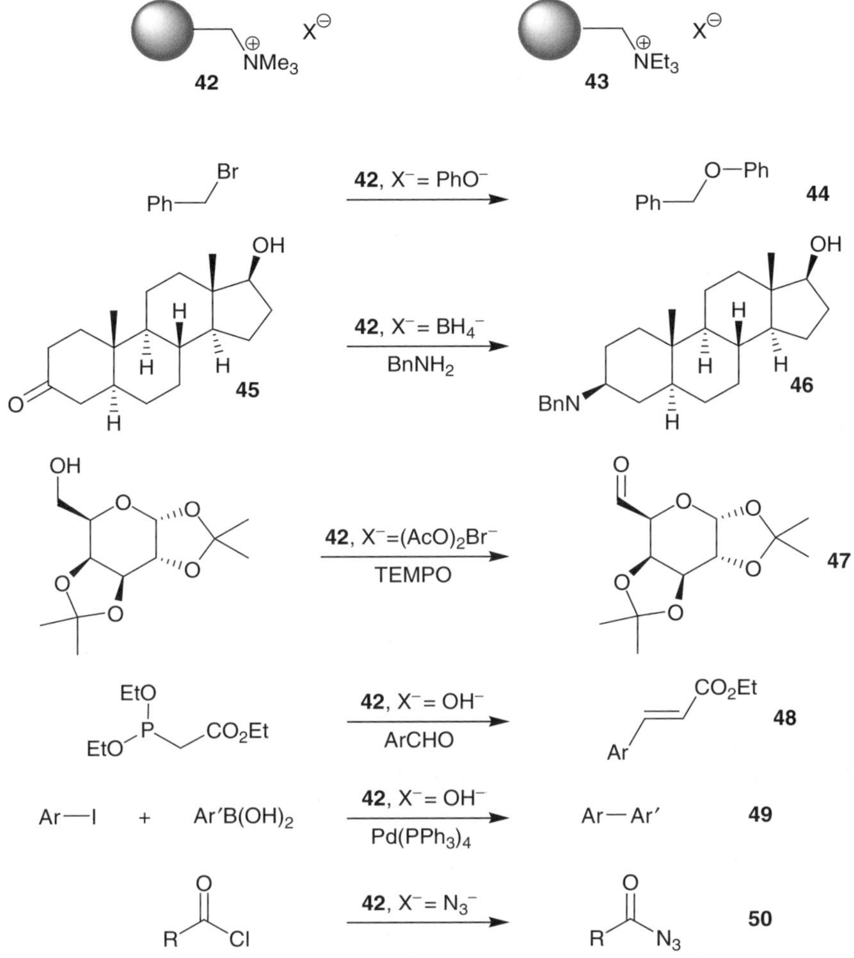

Figure 2.11 Stoichiometric transformations with polymeric monoliths functionalised with quaternary ammonium salts.

rearrangement to isocyanates to afford ureas or carbamates by reaction with different nucleophiles. This not only provided an efficient route to the final products but reduced the handling risks associated with the use of the azide anion.[60] Functionalisation degrees of about $2\,mmol\,g^{-1}$ were obtained, corresponding to cartridge loadings of 3, 7 and 15 mmol [columns with an internal diameter (i.d.) of 100 mm and lengths of 6.6, 10 and 15 cm respectively]. Standard flow rates for these experiments were $0.5\,mL\,min^{-1}$, with residence times of about 13 minutes. The productivity of a single column was up to $30\,mmol\,h^{-1}$.

The functionalisation of monoliths with azlactone (**51**) groups (Figure 2.12) has been used for scavenging, under flow conditions, the excess of amines

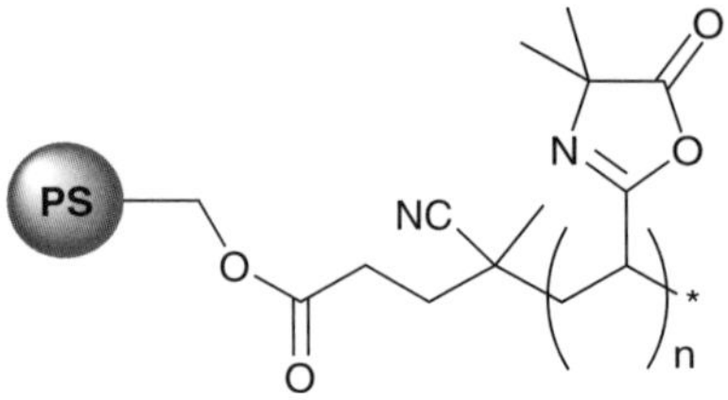

51 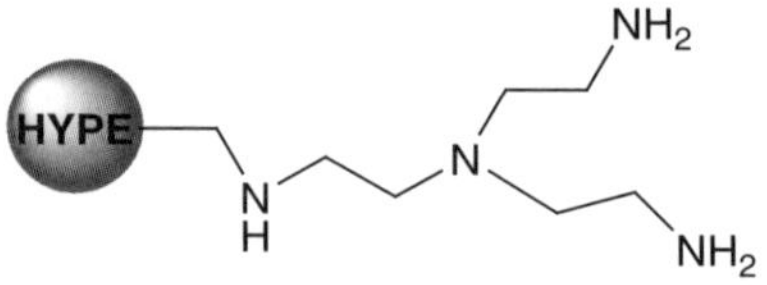   **52**

Figure 2.12 Monolithic scavengers used under flow conditions.

used in a synthesis of ureas.[58] In a similar way, PolyHIPE monoliths derivatised with triethylene tetramines were employed for the removal of acid chlorides.[61]

2.3.3 Monolithic Non-chiral Catalysts

It seems reasonable to assume that catalytic applications are the ones for which full advantage can be taken of the benefits of the combination of supported systems with monolithic materials and flow processes. Thus, it is not surprising that most applications described for polymeric monolithic materials under flow conditions belong to the field of catalytic applications.

Supported onium salts containing a hydroxide counteranion can be applied to a number of different catalytic reactions. One such example is the use of a monolithic PS-DVB supported imidazolium salt as a catalyst for the Henry reaction.[26] Resin-bound imidazolium subunits can be easily introduced onto monolithic polymers through reaction of chloromethyl groups with alkyl imidazol and further methatesis of the initial chloride counteranion to afford the so-called SILLPs (Supported Ionic Liquid Like Phases).[25,26] The use of aqueous solutions of sodium acetate or ammonia for the methatesis reaction provide the access to basic SILLPs (**53**, $R = CH_3$, $X^- = OH^-$, AcO^-) (Figure 2.13). These materials have been used as catalysts for the Henry reaction.

In a flow system, a small amount of substrate is actually forced into intimate contact with an excess of the catalyst. Thus, total conversion can be achieved with residence times as short as 1–3 minutes (flow rates: 0.1–$0.5\,\mathrm{mL\,min}^{-1}$). A small deactivation of the catalytic system is observed for long periods of time, and this seems to be associated with the exchange of the basic anion for other anions formed in the course of the reaction; however, regeneration of the catalyst is easily achieved. The use of nitromethane as both the substrate and the solvent allows work under 'solventless' conditions.

As mentioned above, PASSflow reactors exchanged with the hydroxide anions have been described as bases for the Suzuki reaction, using concomitantly, a Pd(0) catalyst in solution. The preparation of the corresponding Pd(0) supported PASSflow system is also possible using the corresponding chloride forms. These supported palladium species have been shown to be able to catalyse different reactions including:[62,63]

- the transfer hydrogenation to transform benzylic ethers into alcohols or ethyl cinnamate into 3-phenylpropionate;
- the Suzuki reaction, for which a soluble base (KOH) has to be added;
- the Heck reaction between aryl iodides and acrylates using NEt_3 as the base.

Different methodologies were used for the preparation of polymer-supported Pd(0) nanoparticles. The protocol for the preparation of these nanoparticles has a definite impact on their size and distribution.[52,63] The degree of cross-linking and, particularly, the degree of functionalisation with ammonium sites, also have a significant impact on these parameters. Originally, supported palladate species were prepared and reduced to Pd(0) with different agents. The reduction power of those agents also has an influence on the nature of the particles formed. The best results were obtained using sodium borohydride. On the other hand, much smaller nanoparticles were obtained when the loading was carried out under flow conditions (convective loading: 30–55 nm) than when carried out by diffusive loading (44–100 nm).[63]

Comparable results were reported by Ley *et al.* using monoliths derived from PS-DVB polymers functionalised with benzyl triethylammonium groups **55** (Figure 2.13).[64] Initial studies for the Heck reaction showed that a partial leaching of Pd was taking place. Up to 270 parts per million (ppm) of Pd were detected on the solid product after isolation. To avoid this problem, the authors coupled a second column to the exit of the catalytic reactor. This column was

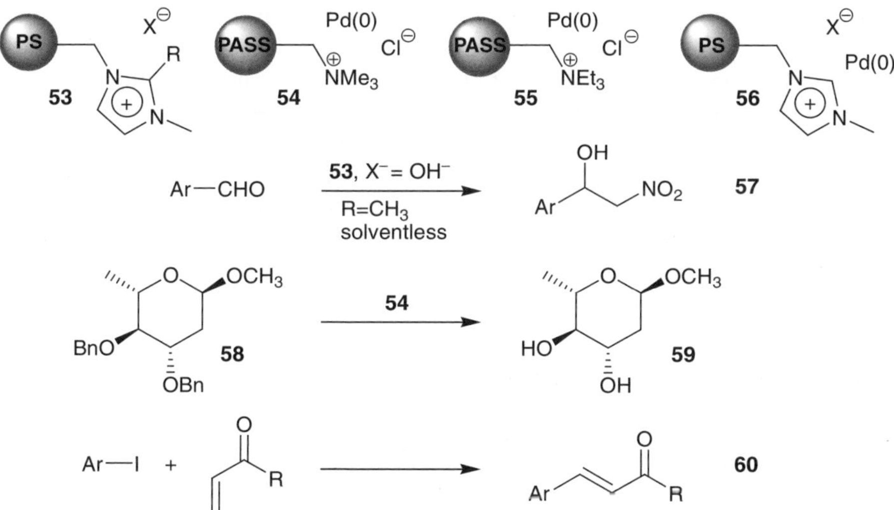

Figure 2.13 Polymer-supported non-chiral catalysts based on onium functionalities.

filled with a commercial polymeric metal scavenger, which allowed the reduction of Pd levels in the final isolated product to below 5 ppm.

These results can be compared with those obtained using Pd(0) containing SILLP **56** (Figure 2.13).[25] The corresponding supported catalysts have been studied for the Heck reaction as well as for the Sonogashira and Suzuki reactions.[65] The ionic liquid (IL) to Pd ratio was a key factor in determining both the nature of the catalytic species formed and the stability of the systems. Working under flow conditions significantly reduces the amount of Pd leaching to the solution—a critical element with which to assess the practical application of this kind of catalytic system. Soluble palladium species seem to participate in the catalytic cycle, but the presence of a high loading of imidazolium groups and the exact nature of those groups on the monolith—in particular the characteristics of the counteranions—determine the ability of the polymer to recapture those Pd species. An appropriate design of the properties of the support, the relative loadings of IL and Pd, and the conditions for the flow process allow the practical suppression of leaching.[25]

As mentioned above, the preparation of monolithic polymers obtained by ROMP allows flow reactors to be obtained containing second generation Grubbs catalysts which have being studied as porous monolithic methatesis catalysts.[54]

Recently, Kirschning *et al.* developed an alternative strategy for the immobilisation of different catalysts on PASSflow reactors.[66] In this case, the catalysts are attached to the solid phase through ionic or coordination bonds. Examples of this methodology include catalysts **61–64** (Figure 2.14), which were prepared using Raschig rings as inert supports. The Ni(0) catalyst **61** was prepared by cation exchange on a sulfonic polymer and further reduction, and has been assayed for the preparation of cyclobutabenzene **65**. The incorporation of pyridine subunits provides an anchoring motif for the immobilisation of other catalytic coordination complexes. Immobilisation of the Pd palladacycle reported in ref. 67 led to Pd(0) catalyst **63**, which has been studied for the Suzuki–Miyaura and related reactions, both in batch and in continuous conditions.[68] Hoveyda and Grubbs-type catalysts **62** and **64** have been studied for ring closing methatesis and related reactions.[69] The main limitation for the use of these catalysts for continuous flow applications is their fast deactivation which, in some instances, is associated with high levels of leaching. Indeed, it has been reported in this case that deactivation under flow conditions can be faster than in batch experiments.[69]

2.3.4 Monolithic Chiral Catalysts

The immobilisation of chiral catalysts on solid polymeric supports probably represents a key application in this field. Besides the abovementioned general advantages, the possibility of recovering and reuse of a catalyst containing a high added value chiral fragment provides a leading direction for the research in this field. The development of efficient flow processes involving such

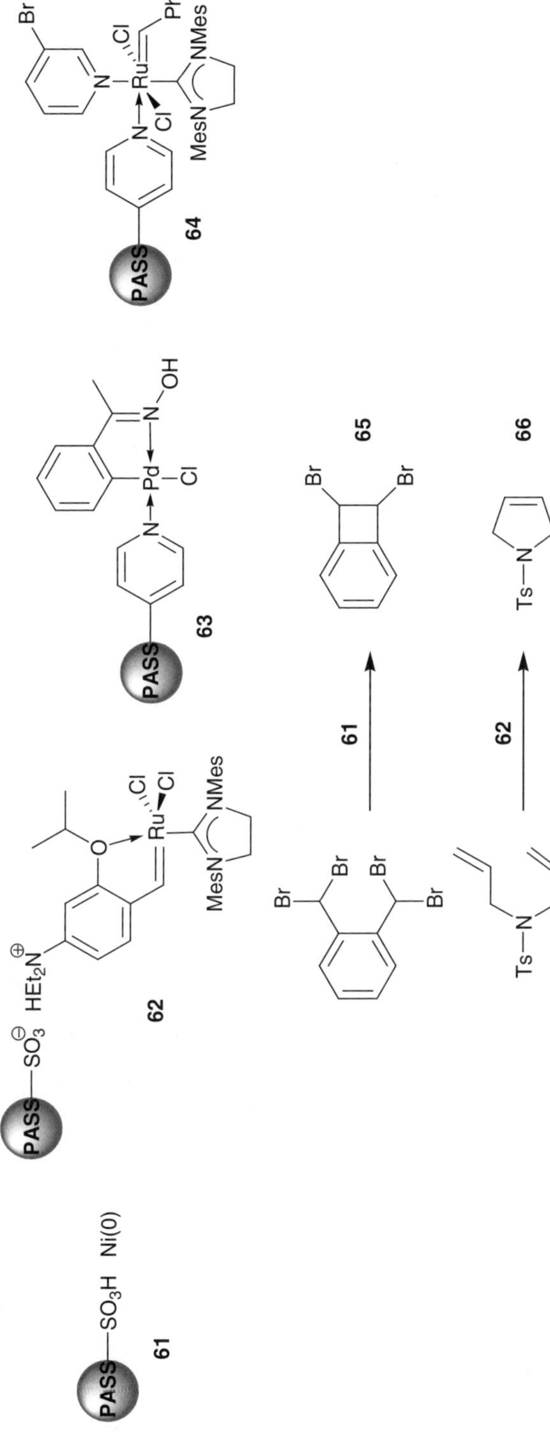

Figure 2.14 PASSflow reactors with ionically or coordinatively bound catalysts.

supported systems would greatly enhance their potential industrial application. For this purpose, monolithic systems seem to be the more appropriate according to their mechanical and fluid dynamic properties.

The first example of the use of asymmetric catalysis based on monolithic materials using flow conditions was reported for the titanium TADDOLates **67**.[70] In this case, the same functional subunit was anchored to different polymeric supports ranging from gel-type resins to monoliths of variable monomeric composition, and the resulting materials were assayed as catalysts for the Diels–Alder reaction between cyclopentadiene and 3-crotonoyl 1,3-oxazolidin-2-one. In the case of the monoliths, the reaction was assayed using a stop-flow approach. One of the most interesting results was the observation that the topicity of the major enantiomer was reversed when the monolithic system was used instead of the gel-type resins or the homogeneous system when Ar = 3,5-dimethyl phenyl. The enantioselectivities found were only moderate (43% ee for Ar = 1-naphthyl), but comparable or even higher than those observed in solution for the analogous soluble catalyst under similar reaction conditions.[31] The monoliths were prepared inside chromatographic stainless steel columns and could be kept active for at least 12–16 months. Regeneration of the spent monolithic columns was not, however, effective. Although the catalysts recovered their activity for the next few cycles, the enantioselectivity of the regenerated systems dropped significantly. Interestingly when the catalyst was used for an alternative reaction under flow conditions, as is the diethyl zinc addition to benzaldehyde, the regenerated column was shown to be as active and enantioselective as a freshly prepared column. Up to 99% ee was obtained using a flow rate of 0.06 mL min^{-1} and a recirculating system working for 2.5 hours.

A second example of an efficient enantioselective catalyst under flow conditions for the addition of diethyl zinc to benzaldehyde is provided by the immobilised chiral aminoalcohol **68** (Figure 2.15).[71] The chiral starting

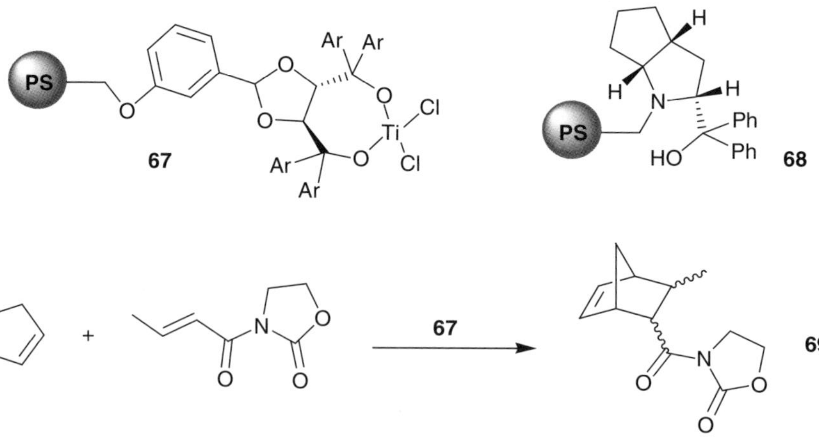

Figure 2.15 Chiral supported catalysts derived from TADDOLs and aminoalcohols.

material for the preparation of **68** is a waste industrial material produced in the synthesis of the ACE inhibitor Ramipril at Aventis.[72] In order to guarantee high conversions, a flow of the reagent and substrate was passed through the monolith, prepared inside a steel column, in recirculation loop mode. The final product was obtained in 85% yield with 99% ee. The same conversion, selectivity and enantioselectivity were maintained for successive cycles and the catalyst did not show any sign of deactivation. The enantioselectivity found under flow conditions was higher than that observed for an analogous system supported on a gel-type resin in batch experiments (up to 89% ee). A similar improvement in enantioselectivity was observed when the comparison was made with the corresponding homogeneous analogue (87% ee under identical conditions). The observation of higher enantioselectivities for flow systems is not uncommon. As in the case of the observation of higher activities, this has been ascribed to the fact that instant catalyst–substrate ratios inside the flow reactor are significantly higher than those under batch conditions.[71]

The immobilisation of bisoxazoline and related chiral ligands on monolithic polymers has also been studied in detail.[73] Monoliths functionalised with bisoxazoline subunits **70**, **71** (Figure 2.16) were prepared through polymerisation of the corresponding functional monomers (containing the chiral fragment) and styrene under a variety of conditions. After conversion into the corresponding copper complexes, the resulting polymers were assayed as catalysts under flow conditions. For this purpose the polymerisation was again carried out inside a stainless steel column so that the connection to a flow system could be carried out in a straightforward way. The benchmark reaction selected in this case was the cyclopropanation of styrene using ethyl diazo acetate (EDA) (Figure 2.16). Significant differences were observed when a comparison was made with the related batch processes;[74] the catalysis under batch conditions showed an activity that was a fifth of that of the homogeneous system. Reaction

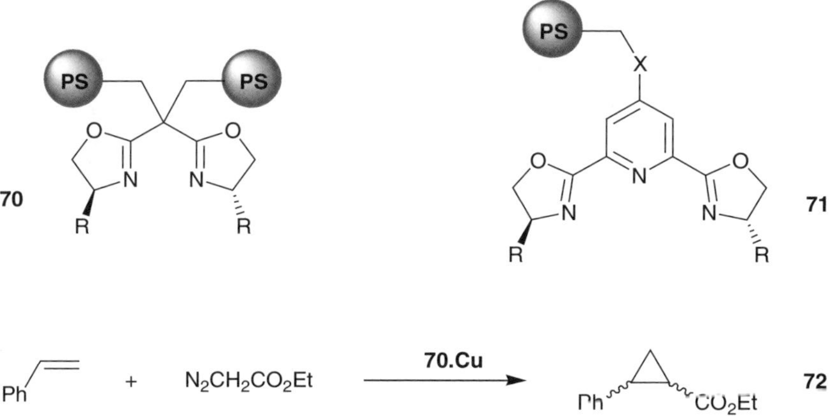

Figure 2.16 Supported bisoxazolines and related systems.

times of more than 16 hours were necessary to obtain an EDA conversion of *ca.* 78%. In the flow system, the conversion was 86% for a residence time of 35 min (flow: $20 \mu l \, min^{-1}$). The conversion could be raised to 100% by reducing the flow to $2 \mu l \, min^{-1}$; in this case the residence time was 350 min. Rather surprisingly, the chemoselectivity was reduced when changing from batch to continuous. Defining the chemoselectivity as cyclopropanes/(cyclopropanes + fumarate + maleate) $\times$ 100, this was 76% for flow experiments; the value for the batch conditions, either homogeneous or with the use of a supported catalyst, was higher than 90%. This cannot be easily rationalised as the chemoselectivity did not change with the different flow rates used and the same kind of functional resin was used for both batch and flow experiments.

The observed enantioselectivity for these systems was very much dependent on the flow rate. The results obtained illustrate nicely the possibilities provided by flow mini-reactors for enantioselective processes. As can be seen in Figure 2.17, an increase in the flow rate is associated with a decrease in enantioselectivity. For the *trans* isomer, the enantiomeric excess values range from 71% at $2 \mu l \, min^{-1}$ to 51% at $200 \mu l \, min^{-1}$. For the *cis* isomer, however, the enantioselectivity remains almost constant at about 55% ee. These results reveal that a significant improvement in enantioselectivity can be obtained in comparison to batch experiments only by correct adjustment of the flow rate. Thus, the enantioselectivities obtained for the homogeneous experiments were 51% (*trans*) and 40% (*cis*), whilst for the reaction using the supported catalyst

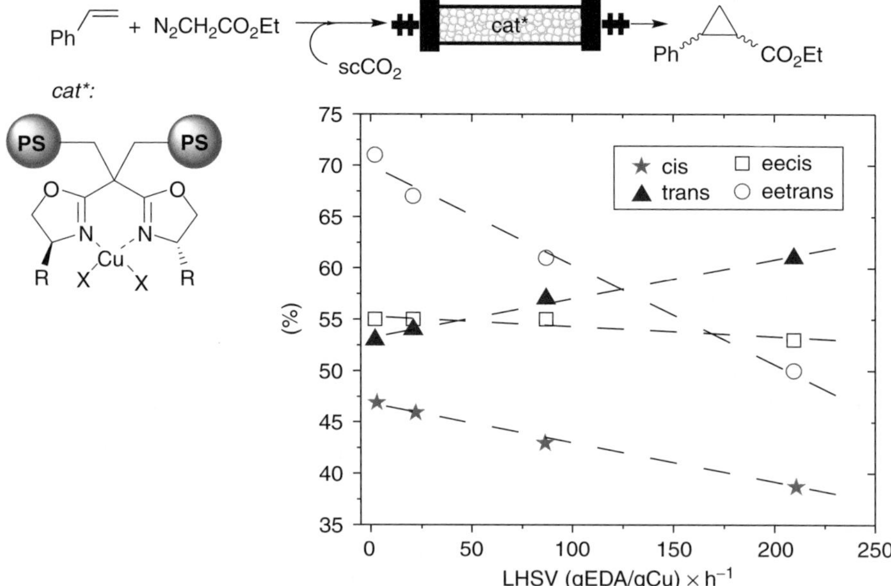

Figure 2.17 Variation of enantioselectivities with flow rates found for supported bisoxazolines and related systems (LHSV: lineal hourly space velocity).

under batch conditions, the values found were 62% (*trans*) and 56% (*cis*). Contrary to general assumptions, the results clearly illustrate that:

- chiral supported catalysts can be more enantioselective than the analogous homogeneous ones;
- flow systems provide additional features to further improve the enantioselectivity through the fine tuning of the reaction conditions such as flow and pressure.

A further benefit of these flow systems was shown to be the possibility of working under solventless conditions and thus avoiding the use of solvents such as DCM with a high environmental impact. For this purpose, the reaction was carried out at $20\,\mu l\,min^{-1}$ with styrene to EDA ratios ranging from 19.2 to 1. The best results were obtained for a styrene to EDA ratio of 4. The enantiomeric excesses were similar to those obtained using DCM under the same flow conditions, but the overall productivity of the system increased three-fold [from 11 to 33 g of cyclopropane $(g\,Cu)^{-1}\,h^{-1}$].

Similar trends were observed when pyridine bis(oxazoline) (pyBOX) supported ligands **71** were used instead of bisoxazolines **70** (Figure 2.16).[75] In this case, the corresponding Ru complexes were used as the catalytic species. An important increase in chemoselectivity was found when the reaction was carried out under solventless conditions. Turn-over frequency (TOF) values increased from 0.3–0.7 moles $(mol\,pyBOX)^{-1}\,h^{-1}$ to 3.3–4.3 moles $(mol\,pyBOX)^{-1}\,h^{-1}$. The enantioselectivities were similar to those found under homogeneous conditions.

An example of chiral PASSflow supported catalyst is given by the Jacobsen-type catalyst **73** (Figure 2.18).[76] This material has been used for efficient dynamic kinetic resolution of racemic epoxides. An example is given by the

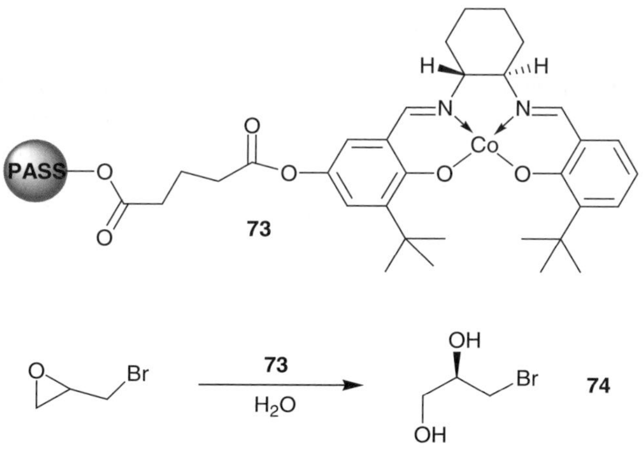

Figure 2.18 Enantioselective dynamic kinetic resolution of racemic epoxides with a PASSflow system.

reaction of bromohydrin in the presence of water, for which a 76% yield was obtained with 91% ee.

2.4 Functionalised Polymers and Potential for Industrial Applications under Flow Conditions

2.4.1 Scaling-up with Polymer-supported Systems

A proper scale-up is critical in order to be able to transform lab-scale processes into industrial applications of practical use. In the case of flow processes using polymer-supported species (stoichiometric or catalytic), three different approaches can be considered:

- scaling-out;
- numbering-up;
- classical scaling-up.

The simplest approach is, obviously scaling-out, *i.e.* the use of the corresponding flow reactor for long periods of time, taking advantage of the stability of these systems and the potential automatisation of the reactor. Ideally, this allows working for 24 hours a day, seven days a week. Nevertheless this approach is rather limited in terms of the final amounts to be obtained. This can be appropriate for high added value products but inadequate for larger productions. In this case, numbering-up or scaling-up (Figure 2.19) is required.

Scaling-up to prepare large flow reactors related to the ones described in Figure 2.5 is not a simple task. Previous experience with ion exchange resins used in large volumes under chromatographic-like conditions has allowed the development of large volume flow reactors based on bead resins. This, however, requires factors such as the mechanical stability of the resins and appropriate packing to eliminate void volumes and preferential channels with uncontrolled fluid dynamics to be taken into consideration. Up to now, these factors have limited this approach to the use of well-known ion exchange bead resins. Scaling-up is much more complicated in the case of monolithic polymers, for which it is difficult to devise an efficient polymerisation process to obtain large-scale monoliths. In this case, but also for bead resins, the numbering-up approach is the most versatile and efficient. The preparation of a multi-tubular reactor with each component having the same dimension and properties of the system at the lab scale is more feasible and, ideally requires only minor adjustments in the reaction or flow conditions.

2.4.2 Use of Ion Exchange Resins as Catalysts for Flow Processes

The most significant contributions to industrial chemistry from the field of functionalised polymers derive from the use of ion exchange resins. Many

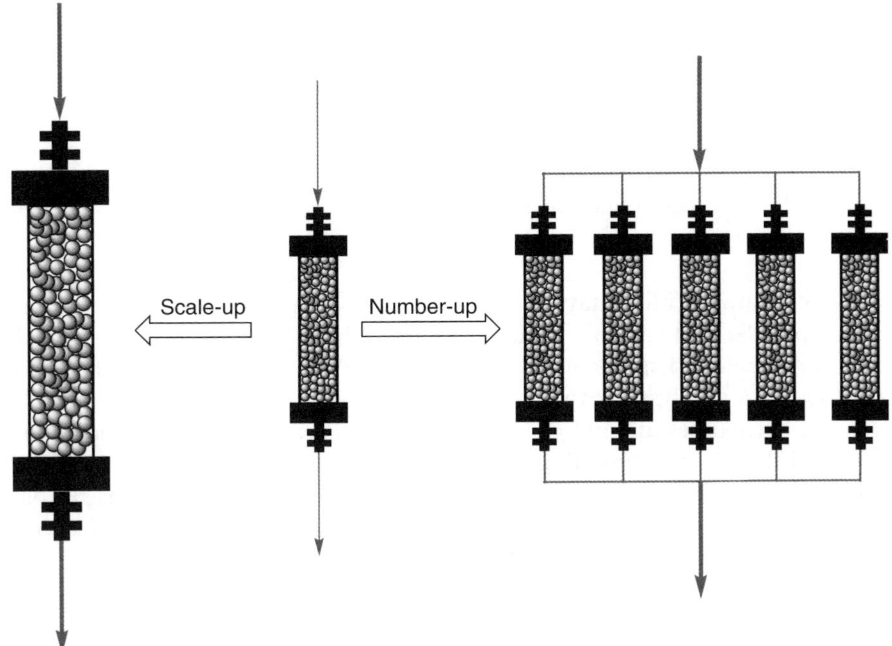

Figure 2.19   Scale-up and numbering-up approaches to large-scale flow reactors.

organic processes are catalysed by strong acids and bases. Thus, it is easy to understand that commercially available ion-exchange resins have been the first functionalised polymers to be used, besides their classical uses, for large-scale laboratory and industrial applications. This field has been reviewed[77] and we will not therefore deal with it here in detail.

Examples of general applications of these kinds of polymers include alkylations, hydrations, etherifications and esterifications, dehydration of alcohols and aldol condensations.[78] The same kind of materials are currently being used as scavengers to remove metals or other impurities from different processes. This application is easily achieved under flow conditions. As most of those polymers are usually macroporous bead resins, the use of type I or type III reactor configurations is most often applied.[79]

Some examples of the preparation of monolithic ion exchange polymers for catalytic applications have also been reported.[80] One example is the preparation of monolithic rods on inert supports that were further sulfonated to produce strong ion exchange polymer composite materials. Their potential was demonstrated for the cyclisation reaction of 1,4-butanediol to tetrahydrofuran using a type III reactor assembly. In this case, the use of small columns with reduced flow rates (1 mL min^{-1}) and pulse injection of the substrate allowed direct coupling of the reaction and separation protocols.

Other recent approaches involve the development of non-classical ion exchange polymers. Those include the study of Nafion and related composite

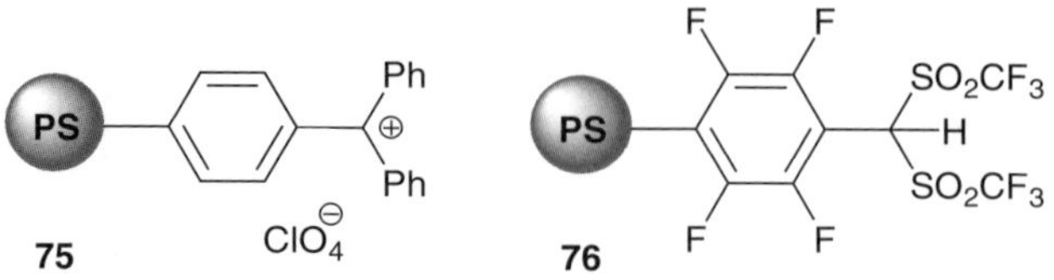

Figure 2.20 Polymer-bound superacidic functionalities.

materials as superacidic catalysts.[81] In addition, acidic functional groups other than carboxylic and sulfonic groups have been introduced in polymeric backbones; Figure 2.20 displays some of these structures. Trityl perchlorate **75** has been used as a catalyst for the condensation of silyl enolethers and aldehydes and some derivatives in good yields.[82] Comparable results were obtained under batch and flow conditions. Polymer-bound Brönsted superacid **76** was developed by Yamamoto and co-workers and used for different reactions, in particular C–C bond forming processes.[83] This polymer was also studied in a type I assembly for the same reactions.[84]

2.5 Ongoing Developments and Future Prospective

2.5.1 Multistage Flow Synthesis with Use of Coupled Columns Packed with Different Functionalised Polymers

The development of packed columns containing polymer-supported reagents, scavengers and catalysts (in bead-type or monolith formats), which can be easily set up in flow systems, opens the way for their sequential assembling to produce multistage chemical reactions. In such processes, each column can play a different role as a catalyst, a reagent or a scavenger.

The most classical example of this concept was provided by Lectka and co-workers,[85] for the synthesis of β-lactams, using three sequentially linked columns.[86] The first column was packed with a polymer-supported base (BEMP resin **19**; Figure 2.7), the second one contained a resin functionalised with a quinine-derived chiral organocatalyst (**77**), and the third one was an amino resin (**78**) to be used as a scavenger (see Figure 2.21). The initial column was fed with an acid chloride (**79**) which was converted *in situ* into a reactive ketene (**80**). An imino ester (**81**) was added to the second column and cyclisation to the β-lactam structure (**82**) took place in the presence of the organocatalyst. The scavenger resin was then used to trap unreacted ketene and imine by-products, affording the expected product in excellent yields and enantioselectivities just by solvent evaporation. The whole flow system was prepared using a simple type I arrangement. Different alternative set-ups of the columns were also assayed, involving for instance the *in situ* generation of the imine through the use of an additional column containing Celite and a basic dehydrogenating agent (sodium hydride). In a third alternative assembly, the polymer **19** and a

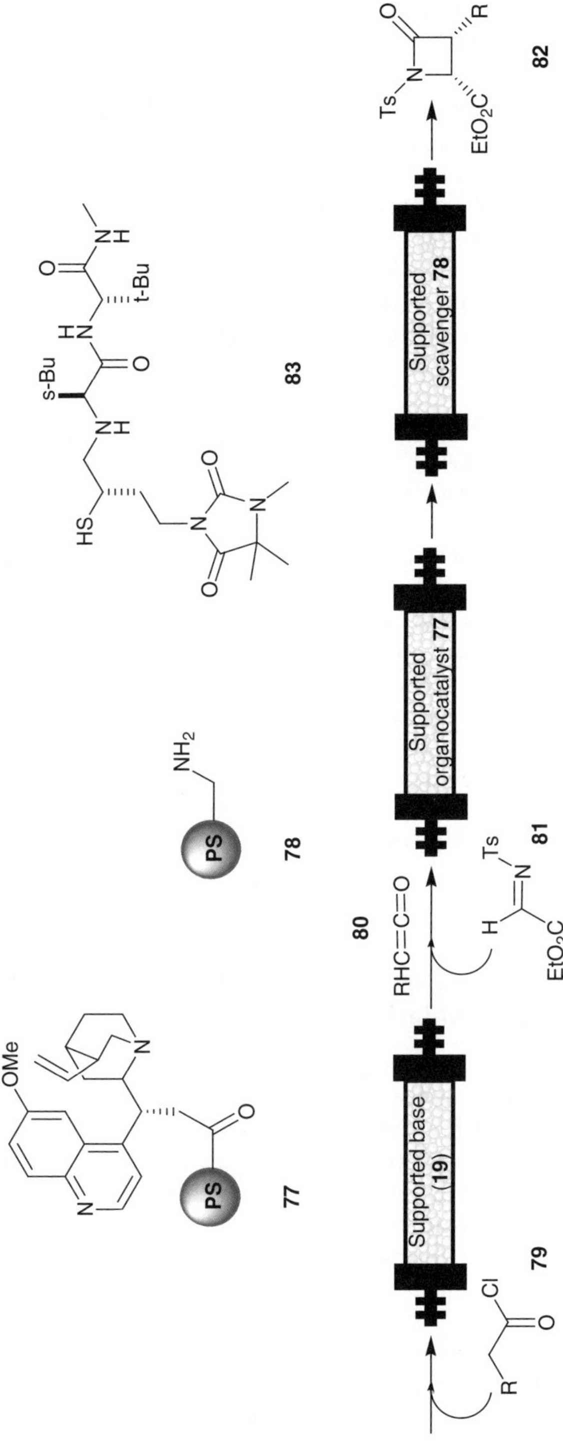

Figure 2.21 Sequential assembly of columns containing different functionalised polymers.

base (K_2CO_3) were mixed in a single column, and the acid chloride and the imine were fed to this column. The same authors then used a related approach to prepare a pharmaceutically active drug candidate. In this case, six different columns containing a combination of reagents, catalysts and scavengers were used in parallel or sequentially to produce the final complex product (**83**) in 34% overall yield (83% ee).[87]

A remarkable example of this approach has been provided by Ley and co-workers in their synthesis of (±)-oxomaritidine **93**, a cytotoxic alkaloid (Figure 2.22).[88] A yield of about 40% (90% purity) was reported after the sequential or parallel use of seven different reactors, five of which were packed with functionalised resins. Three of the resins contained ammonium groups having different counteranions:

- nucleophilic azide (**86**);
- perruthenate as oxidising agent (**87**);[89]
- strongly basic hydroxide (**92**).

The other two resins were a polymer-supported phosphine (**88**) and resin-bound (ditrifluoroacetoxyiodo)benzene (PIFA, **91**)[90] Hydrogenation of a $C=N$ bond on Pd/C (**89**) and an N-acylation step in a microreactor (**90**) were also involved in this synthetic procedure. Scavenging of the unreacted acylating agent was carried out after the microreactor using a small scavenger column (silica-supported primary amine).

Examples with two or three sequentially coupled reactors containing supported reagents, scavengers or catalysts are relatively common. The most frequent examples entail the use of a scavenger resin after a synthetic trans-formation involving the use of a catalytic or stoichiometric functionalised polymer.[91]

2.5.2 Flow Processes Involving Functionalised Polymers and Microwave Irradiation

Development of greener chemical processes can be achieved through the use of different approaches such as the use of catalytic reactions, alternative solvents, solventless conditions with microwave irradiation, process intensification, *etc.* Nevertheless, the goal nowadays is the integration of several of these enabling techniques to facilitate the environmental friendly character of the corre-sponding process.[92] The combination of polymer-supported systems—in par-ticular for catalytic transformations—with flow processes and microwave irradiation clearly represents a step towards this goal.

An example of this approach is the microwave-assisted Suzuki coupling reaction catalysed by palladium-encapsulated catalysts (PdEnCat).[93] The reactor design consists of a glass U-tube packed with the heterogeneous catalyst and the use of relatively slow flow rates (0.1 mL min^{-1}). Maintaining a constant heating power was found to produce an excessive heating of the catalyst and the

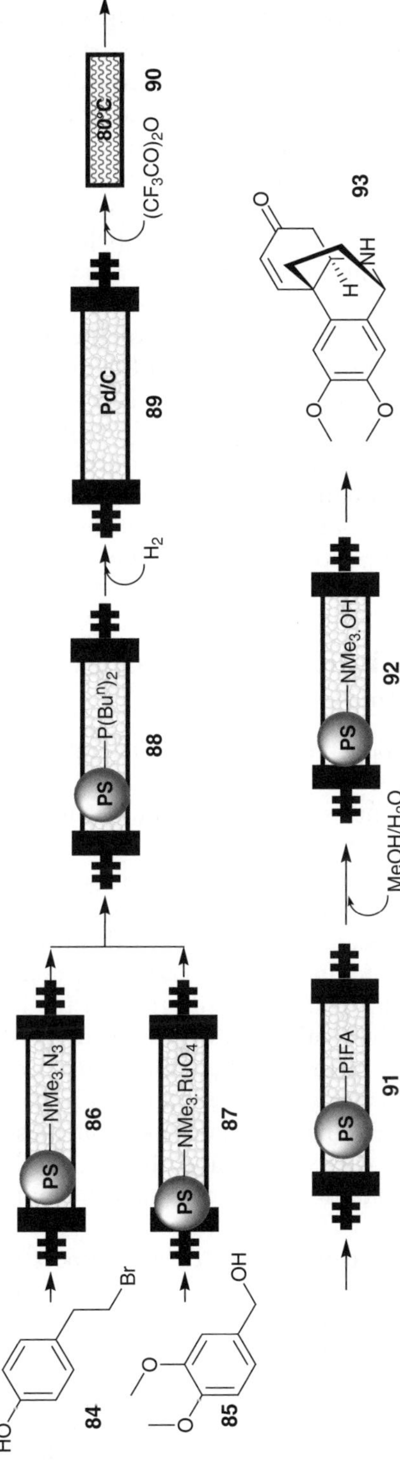

Figure 2.22 Flow multistep synthesis of (±)-oxomaritidine.

collapse of the polymeric matrix.[93] The best results were obtained by a cycling on–off of the heating power. The yields and, in particular, the purities of the resulting products increased (in some cases dramatically) relative to the related batch processes. The amount of product obtained in a single operation represented an 80-fold increase on that produced in a single batch reaction; in addition, the products could be obtained on a multigram scale.

Pd(0) nanoparticles have also been prepared on glass–polymer composite materials in PASSflow set-ups.[63,94] These catalysts have been studied under a variety of conditions, including continuous flow and microwave irradiation, for the transfer hydrogenation of ethyl cinnamate using cyclohexene as the hydrogen donor. Microwave heating profiles for the materials suspended in carbon tetrachloride showed that the presence of the Pd nanoparticles allowed a more efficient conversion of the microwave energy into heat. The results obtained suggest that bigger catalytic particles perform better in microwave heating. In general, the combination of flow processes using heterogeneous systems and microwave irradiation is still in its infancy, and continuous processing under these conditions is a challenge; much more effort is needed to fully develop its potential and to understand the processes taking place.[95]

2.5.3 Flow Processes Involving Functionalised Polymers and Supercritical Fluids

The same considerations abovementioned for the combination with microwave heating are applicable for the combination of flow processes involving functionalized polymers and supercritical fluids. On the other hand, the properties of supercritical fluids (scFs) range between those of gases and liquids. Densities are lower than in the liquid phase, but significantly higher than in the gas phase. Thus, scFs are excellent solvents for a number of substances. As density in this state can be tuned with pressure and temperature changes, dissolution properties may also be varied and adapted. Solvents in the supercritical state possess a smaller viscosity than in the liquid state. The diffusion coefficients of dissolved substances are accordingly high, resulting in a higher transfer rate. Therefore, the use of a scF as a solvent for heterogeneously catalysed chemical reactions may offer the following advantages: enhancement of the reaction rate and selectivity, enhancement of mass and heat transfer rates, an increase in the catalyst lifetime due to the extraction of coke precursors, and easier separation of the products from the solution after reaction.[96]

Different examples of the development of flow processes involving the use of supercritical fluids and polymer-supported catalysts have been reported ranging from Friedel–Crafts alkylation,[97] Suzuki–Miyaura reaction,[98] hydroformilation,[99] aldol reaction,[100] ether synthesis,[101] hydrogenation[102] to "one-pot" reaction involving hydrogenation and aldol condensation.[103] One such example involves the study of polymer supported bisoxazolines and related systems (**70** and **71**). The original solvent for the cyclopropanation reaction was DCM. In order to implement the environmentally friendly character of the process, the same reaction was studied under solventless conditions and with the use of ($scCO_2$) as the solvent. A marked increase in productivity was obtained, in particular with the use of $scCO_2$ (up to 20 times), and essentially maintained unchanged the selectivity and enantioselectivity of the process.[74,75]

2.5.4 Polymer-supported Biocatalysts under Flow Conditions

Biotransformations are key to many industrial applications and, consequently, many continuous flow reactors involving immobilised enzymes are currently in use.[104] We envisage an even broader application of these technologies in the future.

Many different set-ups have been developed for these applications. They involve reactors designs such as the ones shown in Figure 2.5 or the use of membrane reactors.[105] The packed bed reactor (type III reactor), however, is probably the most frequently used assembly. A large variety of supports have also been reported, ranging from inorganic supports to many different organic polymers of polar or non-polar nature.[106]

In the same way, a large variety of methodologies for the immobilisation of enzymes onto functional polymers have been described involving adsorption, entrapment and covalent or ionic attachment. Just a few recent contributions in this field involve the use of microreactors for enzymatic enantioselective C–C bond formation and the use of supported ionic liquids for the immobilisation of enzymes. In the last case, the use of polymers containing imidazolium subunits such as **4** (Scheme 2.1) allowed the immobilisation of lipase B of *Candida antarctica* (CALB). The resulting system was used efficiently for different transformations under flow conditions and with results comparable to those obtained with commercial supported enzymes. An interesting feature of SILLPs-supported enzymes is the increased thermal stability and the compatibility with the use of supercritical fluids.[107] PASSflow systems have also been used for the immobilisation of enzymes.[108]

2.5.5 Miscellaneous Approaches

A number of other strategies are currently being investigated that combine the use of reagents, scavengers and catalysts immobilised on polymers and the development of flow processes. In this chapter we have dealt mainly with polystyrene–divinyl polymers and related materials, but many other polymeric supports have been studied.[11d] A few examples have been mentioned briefly. Those include the use of ROMP matrices,[109] which have been studied by Buchmeiser under flow conditions.[54,55]

On the other hand, the development of microencapsulation and polymer-incarceration methods[110] allow for new approaches to the immobilisation of metal catalysts, and this has been used by Kobayashi and co-workers for the development of flow processes.[111] Thus, a polymer-incarcerated Ru catalyst was used for the selective oxidation of alcohols under batch and flow conditions. The catalyst, mixed with $MgSO_4$, was active under flow conditions (type III assembly; flow: $0.08\,mL\,min^{-1}$) for 8 hours of continuous use with no leaching of Ru being observed and essentially quantitative conversion of the alcohols to aldehydes.

The incorporation of functionalised polymers in the channels of microreactors is another interesting perspective. In one such example, micro-encapsulated catalysts were introduced, modifying the wall channels of the microreactor. The encapsulated Pd catalyst was anchored to the silica walls of the microchannel that had previously been modified with amino groups.[112] The resulting functionalised microreactors were studied for hydrogenation reactions under microflow conditions. Such systems allow an efficient gas–liquid–solid interaction owing to the large interfacial areas and the short path required for molecular diffusion in the very narrow channel space—something which is not attainable in normal batch systems. Thus, very fast and efficient hydrogenation with a mean residence time of 2 minutes can be carried out. Furthermore, in most cases Pd was not detected in the product solutions by inductively coupled plasma (ICP) analysis and the microchannel reactors were reused several times without loss of activity.[113]

In a second example, a microcolumn packed with a catalytic functional resin was integrated in a borosilicate glass micro reactor prepared in-house. The integrated microreactor enabled multicomponent reactions to be performed consisting of both solution phase and heterogeneously catalysed steps.[106] Lewis acid functionalities were present in the corresponding resins and were used as catalysts for the Strecker reaction. Significant operational advantages over batch-wise syntheses were reported:[113]

- faster process (residence time less than 1 minute);
- improved atom efficiency and enhanced yields/purities ($\geq 99.6\%$);
- efficient catalyst recycle;
- safer experimental procedure (minimal exposure to hazardous reactants).

Up to 51 α-aminonitriles were synthesised *via* a series of continuous solution phase and heterogeneously catalysed reaction steps.[114]

On the other hand, Gujit *et al.* have studied the preparation of macroporous monolithic supports for the development of continuous flow capillary microreactors.[115] In this case, the monoliths were prepared *in situ* in narrow bore silica capillaries. The introduction of different ligands and the loading with Pd allow for the preparation of catalyst for C–C coupling reactions that can be used under microfluidic conditions ($0.1\,\mu L\,min^{-1}$) to attain quantitative conversion of the substrates.

An alternative is the development of flow processes based on non-mechanical pumping, in particular on the electroosmotic flow (EOF).[116,117] Here the flow of the mobile phase is produced through the application of an electric field. This strategy has been mainly applied to different designs of microreactors which enable even parallel syntheses to be performed under EOF.[118] The reported advantages are:

- its simplicity;
- the provision of pulse-free reproducible flows;
- the potential to define and alter the magnitude and direction of the flow;
- the minimisation of back pressures.

Furthermore, the ease with which the supported reagents are recycled provides reaction reproducibility and catalyst lifetimes unobtainable in traditional agitated reaction systems. Although in most instances supported species have been immobilised directly on the silica surface of the micro-reactor channels, examples in which the channels are packed with functionalised resins (*i.e.* Amberlite A15, supported piperazine) have also been reported.[119–121]

References

1. J. H. Clark and C. N. Rhodes, *Clean Synthesis Using Porous Inorganic Solid Catalysts and Supported Reagents,* RSC Clean Technology Monographs, Oxford, 2000.
2. A. P. Kybett and D. C. Sherrington, *Supported Catalysts and their Applications,* The Royal Society of Chemistry, Oxford, 2001.
3. A. Kirschning, W. Bannwarth, D. E. Bergbreiter and B. Desai, *Immobilized Catalysts: Solid Phases, Immobilization and Applications,* Springer-Verlag, Berlin and Heidelberg, 2004, Topics in Current Chemistry 242.
4. K. Burgess, *Solid-Phase Organic Synthesis,* John Wiley & Sons, New York, 2000.
5. H. Salimi, A. Rahimi and A. Pourjavadi, *Monatsh. Chem.*, 2007, **138**, 363.
6. P. Seneci, *Solid-Phase Synthesis and Combinatorial Technology,* John Wiley & Sons, New York, 2000.
7. F. Z. Dörwald, *Organic Synthesis on Solid Phase: Supports, Linkers,* Reactions Wiley-VCH, Weinheim, 2002.
8. (a) D. G. Cork and T. Sugawara, *Laboratory Automation in the Chemical Industries,* Marcel Dekker, 2002; (b) M. Koppitz and K. Eis, *Drug Discovery Today*, 2006, **11**, 561.
9. (a) A. Kirschning and G. Jas, *Top. Curr. Chem.*, 2004, **242**, 209; (b) P. Hodge, *Ind. Eng. Chem. Res.*, 2005, **44**, 8542.
10. F. Albericio and J. Tulla-Puche, *The Power of Functional Resins in Organic Synthesis,* Wiley-VCH, Weinheim, 2008.
11. Reviews and specials issues: (a) P. H. Toy and M. Shi, *Polymer-supported Reagents and Catalysts: Increasingly Important Tools for Organic Synthesis, Tetrahedron,* 2005, **61**, 12013; (b) J. Lu, *Chem. Rev.*, 2002, **102**, 3215; (c) T. Kehat, K. Goren and M. Portnoy, *New J. Chem.*, 2007, **31**, 1218; (d) J. Lu and P. H. Toy, *Chem. Rev.*, 2009, **109**, 815.

12. J. W. Labadie, *Curr. Opin. Chem. Biol.*, 1998, **2**, 346.

13. D. C. Sherrington, *Chem Commun.*, 1998, 2275.

14. (a) T. J. Dickerson, N. N. Reed and K. D. Janda, *Chem. Rev.*, 2002, **102**, 3325; (b) D. E. Bergbreiter, J. Tian and C. Hongfa, *Chem. Rev.*, 2009, **109**, 530.

15. (a) P. L. Osburn and D. E. Bergbreiter, *Prog. Polym. Sci.*, 2001, **26**, 2015; (b) D. E. Bergbreiter and S. D. Sung, *Adv. Synth. Catal.*, 2006, **348**, 1352.

16. (a) B. Helms and J. M. J. Fréchet, *Adv. Synth. Catal.*, 2006, **348**, 1125; (b) J. N. H. Reek, S. Arévalo, R. van Heerbeek, P. C. J. Kamer and P. W. N. M. van Leeuwen, *Adv. Catal.*, 2006, **49**, 71.

17. (a) H. P. Dijkstra, G. P. M. van Klink and G. van Koten, *Acc. Chem. Res.*, 2002, **35**, 798; (b) C. Müller, M. G. Nijkamp and D. Vogt, *Eur. J. Inorg. Chem.*, 2005, 4011; (c) E. De Jesús and J. C. Flores, *Ind. Eng. Chem. Res.*, 2008, **47**, 7968.

18. M. L. Dioos, I. F. J. Vankelecom and P. A. Jacobs, *Adv. Synth. Catal.*, 2006, **348**, 1413.

19. M. R. Buchmeiser, *Polymeric Materials in Organic Synthesis and Catalysis,* Wiley-VCH, Weinheim, 2003.

20. F. Svec, T. B. Tennikova and Z. Deyl, *Monolithic Materials: Preparation, Properties and Applications,* Elsevier, Amsterdam, 2003.

21. B. Corain, M. Zecca and K. Jeřábek, *J. Mol. Catal. A: Chem.*, 2001, **177**, 3.

22. (a) P. H. Toy and K. D. Janda, *Tetrahedron Lett.*, 1999, **40**, 6329; (b) P. H. Toy, T. S. Reger, P. Garibay, J. C. Garno, J. A. Malikayil, G. Liu and K. D. Janda, *J. Comb. Chem.*, 2001, **3**, 117; (c) G. Cavalli, A. G. Shooter, D. A. Pears and J. H. G. Steinke, *J. Comb. Chem.*, 2003, **5**, 637.

23. (a) K. Soai and M. Watanabe, *Tetrahedron: Asymmetry*, 1991, **2**, 97; (b) S. V. Luis, M. I. Burguete and B. Altava, *React. Funct. Polym.*, 1995, **26**, 75.

24. (a) V. K Sarin, S. B. H. Kent and R. B. Merrifield, *J. Am. Chem. Soc.*, 1980, **102**, 5463; (b) E. M. Cilli, E. Oliveira, R. Marchetto and C. R. Nakaie, *J. Org. Chem.*, 1996, **61**, 8992; (c) R. Santini, M. C. Griffith and M. Qi, *Tetrahedron Lett.*, 1998, **39**, 8951; (d) A. R. Vaino, D. B. Goodin and K. D. Janda, *J. Comb. Chem.*, 2000, **2**, 330.

25. (a) B. Altava, M. I. Burguete, E. Garcia-Verdugo, N. Karbass, S. V. Luis, A. Puzary and V. Sans, *Tetrahedron Lett.*, 2006, **47**, 2311; (b) N. Karbass, V. Sans, E. Garcia-Verdugo, M. I. Burguete and S. V. Luis, *Chem. Commun.*, 2006, 3095; (c) M. I. Burguete, F. Galindo, E. Garcýa-Verdugo, N. Karbass and S. V. Luis, *Chem. Commun.*, 2007, **29**, 3086.

26. M. I. Burguete, H. Erythropel, E. Garcia-Verdugo, S. V. Luis and V. Sans, *Green Chem.*, 2008, **10**, 401.

27. (a) C. Viklund, F. Svec, J. M. J. Fréchet and K. Irgum, *Chem. Mater.*, 1996, **8**, 744; (b) O. Okay, *Prog. Polym. Sci.*, 2000, **25**, 711; (c) B. P. Santora, M. R. Gagne, K. G. Moloy and N. S. Radu, *Macromolecules*, 2001, **34**, 658; (d) N. Marti, F. Quattrini, A. Butte and M. Morbidelli, *Macromol. Mater. Eng.*, 2005, **290**, 221–229.

28. (a) R. Hahn, M. Panzer, E. Hansen, J. Mollerup and A. Jungbauer, *Sep. Sci. Technol.*, 2002, **37**, 1545; (b) M. Zabka, M. Minceva and A. E. Rodrigues, *J. Biochem. Biophys. Methods*, 2007, **70**, 95; (c) M. K. Danquah and G. M. Forde, *Chem. Eng. J.*, 2008, **140**, 593.

29. (a) R. B. Merrifield, *J. Am. Chem. Soc.*, 1963, **85**, 2149; (b) J. I. Seeman and B. Merrifield, in *Life During a Golden Age of Peptide Chemistry: The Concept and Development of Solid-phase Peptide Synthesis ( )*, American Chemical Society, Washington DC, 1993, Profile, Pathways and Dreams (series).

30. D. C. Sherrington and P. Hogde, *Synthesis and Separations Using Functional Polymers*, Wiley, Chichester, 1988.

31. B. Altava, M. I. Burguete, E. Garcýa-Verdugo, S. V. Luis, M. J. Vicent and J. A. Mayoral, *React. Funct. Polym.*, 2001, **48**, 25.

32. (a) D. W. Kim and D. Y. Chi, *Angew. Chem. Int. Ed.*, 2004, **43**, 483; (b) D. W. Kim, D. J. Hong, K. S. Jang and D. Y. Chi, *Adv. Synth. Catal.*, 2006, **348**, 1719.

33. P. Hodge, *Curr. Opin. Chem. Biol.*, 2003, **7**, 362.

34. (a) A. R. Bogdan, B. P. Mason, K. T. Sylvester and D. T. McQuade, *Angew. Chem. Int. Ed.*, 2007, **46**, 1698; (b) A. R. Bogdan and D. T. McQuade, *Beilstein J. Org. Chem.*, 2009, **5**, 17.

35. S. Itsuno, K. Ito, T. Maruyama, N. Kanda, A. Hirao and S. Nakahama, *Bull. Chem. Soc. Japan.*, 1986, **59**, 3329.

36. K. Kamahori, K. Ito and S. Itsuno, *J. Org. Chem.*, 1996, **61**, 8321.

37. S. Itsuno, Y. Sakurai, K. Ito, T. Maruyama, S. Nakahama and J. M. J. Fréchet, *J. Org. Chem.*, 1990, **55**, 304.

38. P. Hodge, D. W. L. Sung and P. W. Stratford, *J. Chem. Soc., Perkin Trans. I*, 1999, 2335.

39. M. A. Pericàs, C. I. Herrerías and L. Solà, *Adv. Synth. Catal.*, 2008, **350**, 927.

40. N. T. S. Phan, J. Khan and P. Styring, *Tetrahedron*, 2005, **61**, 12065.

41. R. Ballini, L. Barboni, L. Castrica, F. Fringuelli, D. Lanari, F. Pizzo and L. Vaccaro, *Adv. Synth. Catal.*, 2008, **350**, 1218.

42. F. Bonfils, I. Cazaux, P. Hodge and C. Caze, *Org. Biomol. Chem.*, 2006, **4**, 493.

43. (a) S. V. Ley, I. R. Baxendale, R. N. Bream, P. S. Jackson, A. G. Leach, D. A. Longbottom, M. Nesi, J. S. Scott, R. I. Storer and S. J. Taylor, *J. Chem. Soc., Perkin Trans. 1*, 2000, 3815; (b) A. Solinas and M. Taddei, *Synthesis*, 2008, **16**, 2409.

44. R. Harrison and P. Hodge, *J. Chem. Soc., Perkin Trans. 1*, 1976, 2252.

45. R. Harrison and P. Hodge, *J. Chem. Soc., Perkin Trans. 1*, 1982, 509.

46. Mandoli, S. Orlandi, D. Pini and P. Salvadori, *Tetrahedron: Asymmetry*, 2004, **15**, 3233.

47. S. Lundgren, H. Ihre and C. Moberg, *Arkivoc*, 2008, 73.

48. M. R. Buchmeiser, *Polymer*, 2007, **48**, 2187.

49. (a) F. Svec and J. M. J. Frechet, *Science*, 1996, **273**, 205; (b) E. C. Peters, F. Svec and J. M. J. Fréchet, *Adv. Mater.*, 1999, **11**, 1169; (c) F. Svec, *J. Sep. Sci.*, 2004, **27**, 747.

50. Altava, M. I. Burguete, E. García-Verdugo, S. V. Luis and M. J. Vicent, *Green Chem.*, 2006, **8**, 717.

51. N. Hird, I. Hughes, D. Hunter, M. G. J. T. Morrison, D. C. Sherrington and L. Stevenson, *Tetrahedron*, 1999, **55**, 9575.

52. (a) K. Sundmacher, H. Künne and U. Kunz, *Chem. Ing. Tech.*, 1998, **70**, 267; (b) A. Kirschning, C. Altwicker, G. Dräger, J. Harders, N. Hoffmann, U. Hoffmann, H. Schönfeld, W. Solodenko and U. Kunz, *Angew. Chem. Int. Ed.*, 2001, **40**, 3995; (c) U. Kunz, A. Kirschning, H.-L. Wen, W. Solodenko, R. Cecilia, C. O. Kappe and T. Turek, *Catal. Today*, 2005, **105**, 318.

53. J. F. Brown, P. Krajnc and N. R. Cameron, *Ind. Eng. Chem. Res.*, 2005, **44**, 8565.

54. (a) M. R. Buchmeiser, *Angew. Chem. Int. Ed.*, 2001, **40**, 3795; (b) M. Mayr, B. Mayr and M. R. Buchmeiser, *Angew. Chem. Int. Ed.*, 2001, **40**, 3839; (c) F. Sinner and M. R. Buchmeiser, *Macromolecules*, 2000, **33**, 5777; (d) F. Sinner and M. R. Buchmeiser, *Angew. Chem. Int. Ed.*, 2000, **39**, 1433.

55. S. Lubbad, B. Mayr, M. Mayr and M. R. Buchmeiser, *Macromol. Symp.*, 2004, **210**, 1.

56. S. Zalusky, R. Olayo-Valles, C. J. Taylor and M. A. Hillmyer, *J. Am. Chem. Soc.*, 2001, **123**, 1519.

57. J. A. Tripp, F. Svec and J. M. J. Fréchet, *J. Comb. Chem.*, 2001, **3**, 604.

58. (a) J. A. Tripp, J. A. Stein, F. Svec and J. M. J. Fréchet, *J. Org. Lett.*, 2000, **2**, 195; (b) J. A. Tripp, F. Svec and J. M. J. Fréchet, *J. Comb. Chem.*, 2001, **3**, 216.

59. (a) G. Sourkouni-Argirusi and A. Kirschning, *Org. Lett.*, 2000, **2**, 3781; (b) W. Solodenko, U. Kunz and A. Kirschning, *Bioorg. Med. Chem. Lett.*, 2002, **12**, 1833; (c) U. Kunz, H. Schönfeld, A. Kirschning and W. Solodenko, *J. Chromatogr. A*, 2003, **1006**, 241; (d) M. Brünjes, G. Sourkouni-Argirusi and A. Kirschning, *Adv. Synth. Catal.*, 2003, **345**, 635.

60. M. Baumann, I. R. Baxendale, S. V. Ley, N. Nikbin and C. D. Smith, *Org. Biomol. Chem.*, 2008, **6**, 1587.

61. P. Krajnc, J. F. Brown and N. R. Cameron, *Org. Lett.*, 2002, **4**, 2497.

62. W. Sodolenko, H. Wen, S. Leue, F. Stuhlmann, G. Sourkouni-Argirusi, G. Jas, H. Schönfeld, U. Kunz and A. Kirschning, *Eur. J. Org. Chem.*, 2004, 3601.

63. K. Mennecke, R. Cecilia, T. N. Glasnov, S. Gruhl, C. Vogt, A. Feldhoff, M. A. Larrubia Vargas, C. O. Kappe, U. Kunz and A. Kirschning, *Adv. Synth. Catal.*, 2008, **350**, 717.

64. N. Nikbin, M. Ladlow and S. V. Ley, *Org. Process Res. Dev.*, 2007, **11**, 458.

65. S.V. Luis and E. Garcia-Verdugo, unpublished results.

66. (a) A. Kirschning, W. Solodenko and K. Mennecke, *Chem. Eur. J.*, 2006, **12**, 5972; (b) K. Mannecke and A. Kirschning, *Synthesis*, 2008, 3267; (c) K. Mennecke, W. Solodenko and A. Kirschning, *Synthesis*, 2008, **1**, 1589.

67. D. A. Alonso, C. Najera and M. C. Pacheco, *Org. Lett.*, 2000, **2**, 1823.

68. W. Solodenko, K. Mennecke, C. Vogt, S. Gruhl and A. Kirschning, *Synthesis*, 2006, 1.

69. (a) K. Mennecke, K. Grela, U. Kunz and A. Kirschning, *Synlett*, 2005, 2948; (b) A. Michrowska, K. Mennecke, U. Kunz, A. Kirschning and K. Grela, *J. Am. Chem. Soc.*, 2006, **128**, 13261.

70. Altava, M. I. Burguete, J. M. Fraile, J. I. GarcIa, S. V. Luis, J. A. Mayoral and M. J. Vicent, *Angew. Chem. Int. Ed.*, 2000, **39**, 1503.

71. M. I. Burgurette, E. Garcia-Verdugo, M. J. Vicent, S. V. Luis, H. Pennemann, N. G. von Keyserling and J. Martens, *Org. Lett.*, 2002, **4**, 3947.

72. A. Kleemann, J. Engel, B. Kutscher and D. Reichert, *Pharmaceutical Substances*, Thieme, Stuttgart, 4th edn, 2001, p. 1785.

73. (a) G. Desimoni, G. Faita and K. A. Jorgensen, *Chem. Rev.*, 2006, **106**, 3561; (b) R. Rasappan, D. Laventine and O. Reiser, *Coord. Chem. Rev.*, 2008, **252**, 702.

74. M. I. Burguete, A. Cornejo, E. García-Verdugo, J. García, M. J. Gil, S. V. Luis, V. Martínez-Merino, J. A. Mayoral and M. Sokolova, *Green Chem.*, 2007, **9**, 1091.

75. M. I. Burguete, A. Cornejo, E. Garca-Verdugo, M. J. Gil, S. V. Luis, J. A. Mayoral, V. Martinez-Merino and M. Sokolova, *J. Org. Chem.*, 2007, **72**, 4344.

76. W. Solodenko, G. Jas, U. Kunz and A. Kirschning, *Synthesis*, 2007, 583.

77. G. Gelbard, *Ind. Eng. Chem. Res.*, 2005, **44**, 8468.

78. (a) H. Widdecke, *Br. Polym. J.*, 1984, **16**, 188; (b) M. M. Sharma, *React. Funct. Polym.*, 1995, **26**, 3; (c) M. A. Harmer and Q. Sun, *Appl. Catal. A*, 2001, **221**, 45; (d) K. Tanabe and W. F. Hölderich, *Appl. Catal. A*, 1999, **181**, 399.

79. E. Garrett and K. Prasad, *Adv. Synth. Catal.*, 2004, **346**, 889.

80. U. Kunz, C. Altwicker, U. Limbeck and U. Hoffmann, *J. Mol. Catal. A: Chem.*, 2001, **177**, 21.

81. Á. Molnár, *Curr. Org. Chem.*, 2008, **12**, 159.

82. T. Mukaiyama and H. Iwakiri, *Chem. Lett.*, 1985, 1363.

83. K. Ishiara, A. Hasegawa and H. Yamamoto, *Angew. Chem. Int. Ed.*, 2001, **40**, 4077.

84. K. Ishiara, A. Hasegawa and H. Yamamoto, *Synlett*, 2002, 1296.

85. M. Hafez, A. E. Taggi and T. Lectka, *Chem. Eur. J.*, 2002, **8**, 4115.

86. (a) A. M. Hafez, A. E. Taggi, H. Wack, W. J. Drury and T. Lectka, *Org. Lett.*, 2000, **2**, 3963; (b) A. M. Hafez, A. E. Taggi, T. Dudding and T. Lectka, *J. Am. Chem. Soc.*, 2001, **123**, 10853.

87. S. France, D. Bernstein, A. Weatherwax and T. Lectka, *Org. Lett.*, 2005, **7**, 3009.

88. R. Baxendale, J. Deeley, C. M. Griffiths-Jones, S. V. Ley, S. Saaby and G. K. Tranmer, *Chem. Commun.*, 2006, 2566.

89. (a) B. Hinzen and S. V. Ley, *J. Chem. Soc., Perkin Trans. 1*, 1997, 1907; (b) R. Lenz and S. V. Ley, *J. Chem. Soc., Perkin Trans. 1*, 1997, 3291; (c) B. Hinzen, R. Lenz and S. V. Ley, *Synthesis*, 1998, 977.

90. (a) S. V. Ley, A. W. Thomas and H. Finch, *J. Chem. Soc., Perkin Trans. 1*, 1999, 669; (b) H. Toga, G. Nogami and M. Yokoyama, *Synlett*, 1998, 534.

91. (a) I. R. Baxendale, S. V. Ley, C. D. Smith and G. K. Tranmer, *Chem. Commun.*, 2006, 4835; (b) J. Sedelmeier, S. V. Ley and I. R. Baxendale, *Green Chem.*, 2009, **11**, 683.

92. E. Bergbreiter and S. Kobayashi, *Chem. Rev.*, 2009, **109**, 257.

93. R. Baxendale, C. M. Griffiths-Jones, S. V. Ley and G. K. Tranmer, *Chem. Eur. J.*, 2006, **12**, 4407.

94. R. Cecilia, U. Kunz and T. Turek, *Chem. Eng. Proc.*, 2007, **47**, 870.

95. Leadbutter, Organic Process and Research Development 2007.

96. For reviews of heterogeneous catalysis in scCO$_2$, see: (a) A. Baiker, *Chem. Rev.*, 1999, **99**, 453; (b) P. G. Jessop and W. Leitner, *Chemical Synthesis Using Supercritical Fluids*, Wiley-VCH, Weinheim, 1999; (c) J.-D. Grunwaldt, R. Wandeler and A. Baiker, *Catal. Rev. Sci. Eng.*, 2003, **45**, 1; (d) J.-D. Grunwaldt and A. Baiker, *Phys. Chem. Chem. Phys.*, 2005, **7**, 3526.

97. R. Amandi, P. Licence, S. K. Ross, O. Aaltonen and M. Poliakoff, *Org. Process. Res. Dev.*, 2005, **9**, 451.

98. G. A. Leeke, R. C. D. Santos, B. Al-Duri, J. P. K. Seville, C. J. Smith, C. K. Y. Lee, A. B. Holmes and I. F. McConvey, *Org. Process Res. Dev.*, 2007, **11**, 144.

99. F. Shibahara, K. Nozaki and T. Hiyama, *J. Am. Chem. Soc.*, 2003, **125**, 8555.

100. G. Stevens, R. A. Bourne and M. Poliakoff, *Green Chem.*, 2009, **11**, 409.

101. J.-K. Lee, M. J. Fuchter, R. M. Williamson, G. A. Leeke, E. J. Bush, I. F. McConvey, S. Saubern, J. H. Ryan and A. B. Holmes, *Chem. Commun.*, 2008, 4780.

102. T. Seki, J.-D. Grunwaldt and A. Baiker, *Ind. Eng. Chem. Res.*, 2008, **47**, 4561.

103. (a) T. Seki, J.-D. Grunwaldt and Alfons Baiker, *Chem. Commun.*, 2007, 3562; (b) T. Seki, J.-D. Grunwaldt, N. van Vegten and A. Baiker, *Adv. Synth. Catal.*, 2008, **350**, 691.

104. N. End and K.-U. Schöning, *Top. Curr. Chem.*, 2004, **242**, 273.

105. N. N. Rao, S. Lütz, K. Würges and D. Minör, *Org. Process Res. Dev.*, 2009, **13**, 607.

106. R. A. Sheldon, *Adv. Synth. Catal.*, 2007, **349**, 1289.

107. P. Lozano, E. García-Verdugo, R. Piamtongkam, N. Karbass, T. De Diego, M. I. Burguete, S. V. Luis and J. L. Iborra, *Adv. Synth. Catal.*, 2007, **349**, 1077.

108. G. Dräger, C. Kiss, U. Kunz and A. Kirschning, *Org. Biomol. Chem.*, 2007, **5**, 3657.

109. N. Madhavan, C. W. Jones and M. Weck, *Acc. Chem. Res.*, 2008, **41**, 1153.

110. R. Akiyama and S. Kobayashi, *Chem. Rev.*, 2009, **109**, 594.

111. S. Kobayashi, H. Miyamura, R. Akiyama and T. Ishida, *J. Am. Chem. Soc.*, 2005, **127**, 9251.

112. J. Kobayashi, Y. Mori, K. Okamoto, R. Akiyama, M. Ueno, T. Kitamori and S. Kobayashi, *Science*, 2004, **304**, 1305.

113. C. Wiles and P. Watts, *Org. Process Res. Dev.*, 2008, **12**, 1001.
114. C. Wiles and P. Watts, *Eur. J. Org. Chem.*, 2008, 5597.
115. (a) K. F. Bolton, A. J. Canty, J. A. Deverell, R. M. Guijt, E. F. Hilder, T. Rodemann and J. A. Smith, *Tetrahedron Lett.*, 2006, **47**, 9321; (b) A. J. Canty, J. A. Deverell, A. Gömann, R. M. Guijt, T. Rodemann and J. A. Smith, *Aust. J. Chem.*, 2008, **61**, 630; (c) A. Gömann, J. A. Deverell, K. F. Munting, R. C. Jones, T. Rodemann, A. J. Canty, J. A. Smith and R. M. Guijt, *Tetrahedron*, 2009, **65**, 1450.
116. C. Wiles, P. Watts, S. J. Haswell and E. Pombo-Villar, *Tetrahedron*, 2005, **61**, 10757.
117. C. Wiles, P. Watts and S. J. Haswell, *Chem. Commun.*, 2007, 966.
118. C. Wiles and P. Watts, *Chem. Commun.*, 2007, 4928.
119. C. Wiles, P. Watts and S. J. Haswell, *Tetrahedron*, 2005, **61**, 5209.
120. C. Wiles, P. Watts and S. J. Haswell, *Lab Chip*, 2007, **7**, 322.
121. N. Nikbin and P. Watts, *Org. Process Res. Dev.*, 2004, **8**, 942.

Zeolites and Related Materials for Developing Continuous Flow Systems

MARIA J. SABATER,[a] FERNANDO REY[a] AND JESÚS LÁZARO[b]

[a] Instituto de Tecnología Química, UPV-CSIC, Universidad Politécnica de Valencia, Avenida los Naranjos s/n, 46022, Valencia, Spain; [b] Compañía Española de Petróleos. Centro de Investigación, 28850, San Fernando de Henares, Madrid, Spain

3.1   Introduction

Nowadays, a high percentage of chemical products manufactured in industry are obtained through catalytic procedures. Indeed dehydration, condensation, isomerisation, alkylation, etherification, amination, cracking, esterification reactions, *etc.* are reported to be catalysed by solid acid–base catalysts; more than 40% of these mentioned processes are catalysed by zeolites.[1]

Zeolites have received most attention in industrial applications and have contributed to the great progress made by the petroleum and petrochemical industry for the last decades.[2,3] They have shown a long-established tradition in bulk chemicals and oil refining, where continuous flow processes are well implemented in dedicated plants. In sharp contrast, zeolites have been less exploited in the synthesis of building block chemicals, where reactions are generally performed in the liquid phase in multipurpose equipment.

RSC Green Chemistry No. 5
Chemical Reactions and Processes under Flow Conditions
Edited by S.V. Luis and E. Garcia-Verdugo

Published by the Royal Society of Chemistry, www.rsc.org

In this chapter, we highlight the most important applications of zeolite catalysts in the refining and petrochemical industries as well as in the fine chemicals industry. One major section deals with zeolite applications in petroleum refining and petrochemistry, where classic examples of continuous flow processes such as catalytic cracking, isomerisation of light gasoline and light alkenes, *etc.* are addressed. As well as these traditional areas, new applications of zeolite catalysts (*e.g.* in the manufacture of organic intermediates and fine chemicals) are steadily emerging and are addressed in a separate section.

Although zeolites are increasingly utilised in other areas as multifunctional materials such as sensor or membrane technology, their main application (in economic terms) still lies in heterogeneous catalysis.

3.2 Zeolites and Zeotypes: Outstanding Inorganic Materials for Heterogeneous Processes in Chemistry

Zeolites are tridimensional crystalline aluminosilicates with the following formula in the 'as-synthesised' form:

$$xM_{2/n}O \cdot xT'_2O_3 \cdot yTO_2 \cdot wH_2O$$

where: M is the cation, which can belong to group IA or IIA or can be an organic cation; n is the cation valence; T and T' are a tetravalent and a trivalent cation respectively; and w represents the number of water molecules contained in the zeolite voids.

In these crystalline structures, T and T' occupy framework positions in tetrahedral coordination with oxygen, and typically are Si and Al [SiO_4^{4-} and AlO_4^{5-} (Figure 3.1)].[4]

A zeolite structure containing only SiO_2 tetrahedra would be electrically neutral and no acidity would be developed on its surface. Brönsted sites are developed when Si^{4+} is isomorphically substituted by a trivalent cation (*e.g.* Al^{3+}) and a negative charge is created in the lattice, which is compensated for by a proton. The proton is attached to the oxygen atom connected to neighbouring silicon and aluminium atoms, resulting in the so-called bridged

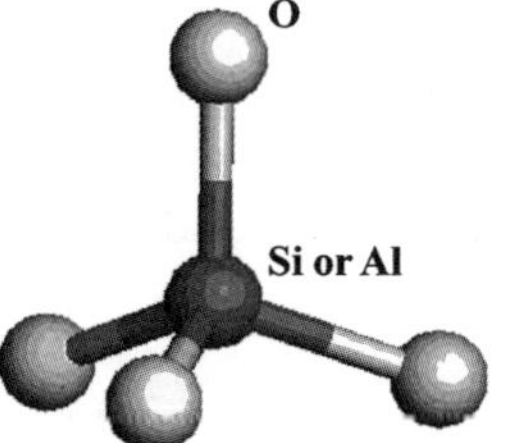

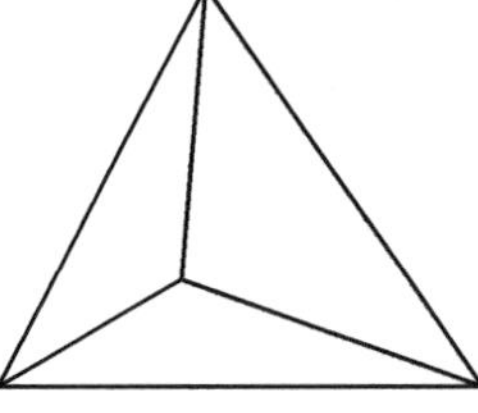

Figure 3.1 Representations of SiO_4^{4-} and AlO_4^{5-} tetrahedra.

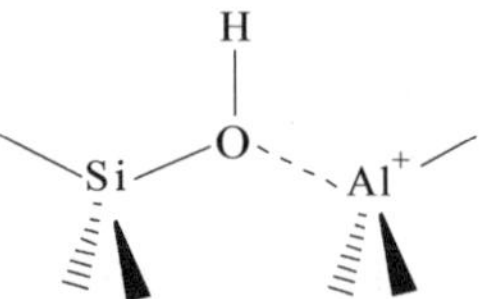

Figure 3.2 Model for the chemical structure of the bridged hydroxyl in zeolites.

hydroxyl group which is the site responsible for the Brönsted acidity of zeolites (Figure 3.2).[5,6]

Other microporous structures have been synthesised in which not only Si and Al, but also P, transition metals and many other groups elements with a valence ranging from I to V such as B, Ga, Fe, Cr, Ti, V, Mn, Co, Zn, Cu, *etc.* occupy framework positions. These are reported with the generic name of zeotypes including $ALPO_4$, SAPO, MeAPO and MeAPSO type molecular sieves.[7-13]

The main characteristic of these zeolitic materials is their crystalline stable structure enclosing a well-defined system of pores, which are conventionally defined as ultralarge. The pores are formed with more than 12 member ring (MR) pore apertures, (*i.e.* the largest channel is delimited by more than 12 tetrahedral atoms), large (12MR), medium (10MR) and small (8MR) pore zeolites, and whose diameter can vary between 5 and 20 Å. Besides this, there are zeolites with pore apertures smaller than 8MR (generally named as clathrasils), but their inner space is not accessible for any molecule and therefore, they do not have any catalytic interest.

The porous network of these materials means that they present a high surface area which is able to adsorb very large quantities of hydrocarbons. This fact combined with the possibility of generating active sites inside the channels and cavities produces a very unique type of catalyst, which itself can be considered as a catalytic nanoreactor.

It is well-recognised that the success of zeolites in catalysis relies on their well-defined pore dimensions and topological connectivities, which provide the potential to produce shape selectivity in many reactions, as well as in their cation exchange ability.

It is important to note that the acid–base properties of these materials can be controlled by:

- selecting the type of ion exchanged cation;
- the Si/Al ratio of the zeolite framework;
- the nature of the tetravalent (Si, Ge, Ti, Sn, *etc.*) and trivalent (Al, B, Ga, *etc.*) cations forming the zeolite network.

Nonetheless, it is necessary to take into account that a wide variation of acid–base properties can be achieved by ion exchange and ion addition, while a relatively small change in acid–base properties is yielded by changing the Si/Al ratio.[5]

With respect to basicity, two approaches have been undertaken to prepare basic zeolites. One approach is ion exchange with alkali metal ions, and the other is to impregnate the zeolite pores with fine particles that can act as bases themselves. The former produces relatively weak basic sites, while the latter results in strong basic sites.[14]

With alkali ion exchanged zeolites, the type of alkali used affects the basic strength of the resulting zeolites. Effects of the alkali ion on basic strength are in the following order: $Cs^+ > Rb^+ > K^+ > Na^+ > Li^+$. Their basic sites are framework oxygen. The bonding of the framework oxygen is rather covalent in nature. This causes the basic sites of ion exchanged zeolites to be relatively weak compared to, for example, those of alkaline earth oxides.[14]

The zeolitic structure (especially that of alkali ion-added zeolites) often collapses during preparation procedures. In close connection to this, Yagi *et al.* prepared Cs ion-added zeolites to establish the preparative conditions necessary to retain the zeolite framework during preparative procedures.[14d] It was found that the crystalline structures of zeolites, in particular alkali ion-added zeolites, are easily destroyed by exposure to water vapour at high temperatures and that zeolites of high Si/Al ratio are unstable to alkali treatment.

In conclusion, zeolites offer a family of acid–base materials with the potential for extremely fine tuning of the acid–base properties as well as the adsorption properties. This provides an extremely efficient family of catalysts able to maximise the selectivity of many industrial processes towards the target product while reducing the formation of undesired secondary products. However, it must be said that the main benefit of zeolites (*i.e.* the very narrow pore distribution within the micropore diameter) is also their main drawback, since the processing of bulky molecules cannot take place using these microporous solids because the reactants cannot reach the active sites located inside the zeolite pores. The advent of mesoporous ordered materials was able to cover this gap, but their thermal and hydrothermal stability needs to be improved to allow industrial application.[15] Thus, the pursuit of extra large pore zeolites remains of great interest for chemical industries.[16]

In summary, more than 170 distinct zeolite and molecular sieve structure types are recognised today by the International Zeolite Association (IZA).[4,17] The number has increased steadily over the last 30 years. In addition, the number of patents issued for the composition or use of zeolites and molecular sieves continues to growth, indicating a clear industrial interest in the utility of these materials.[4]

However, despite the continued increase in availability of new molecular sieves and the associated research activities into their application, only a few are used in commercial processes. The discrepancy between available materials and commercial useful products can be understood in the overall context of the requirements for new technology. For example, any porous catalyst particle must meet the demands both of flow processes (*vs.* pressure drop in fixed bed catalyst) or of viability of separation (suspended catalyst particles in batch reactions).

In addition, most of these materials are used as part of a macro-composite blend of ingredients that typically includes other process-active reagents of some type. Therefore this new catalytic material must not only be able to survive and perform under the specific operation regimes of the application under consideration, but it must also be compatible with other reaction components or reaction conditions.

Likewise, the performance of this material must provide an economic benefit that justifies the costs invested in both the synthesis of the material and/or its modification in order to justify its commercial use.

3.3 Current Industrial Applications of Zeolites and Related Materials

Solid acid catalysts have served as important functional materials for the petroleum refinery industry and for the production of chemicals. At present, however, a significant number of acid-catalysed reactions (Friedel–Crafts reactions, esterifications, hydrolysis) are still carried out using conventional acids such as H_2SO_4 and $AlCl_3$. These processes are associated with problems of high toxicity, corrosion, catalyst waste, difficulty of separation and recovery. Thus replacement of these liquids acids with solid acids is desirable in the chemical industry.

Zeolites, with their numerous adjustable catalytic properties, offer a broad range of possibilities for industrial applications. Among these industrial processes, several are of significant importance and actually operate at industrial scale. These are described in more detail below.

3.3.1 Zeolites in Refining and Petrochemical Processes

There is no doubt that nowadays the employment of zeolites as solid acid catalysts is one of the cornerstones in the production of liquid fuels in refineries.[18] Indeed, as described in this chapter, many commercial processes in refineries are based on zeolitic catalysts.

There is no single reason for this wide use of zeolites in refinery and petrochemical processes, but many reasons converge for their fruitful employment in petroleum chemistry. Perhaps the most important are those listed below:

- The thermal and hydrothermal stability of zeolites allows their use under severe conditions—as generally found in refineries. This property is particularly important during the regeneration of the acid properties of the catalysts, which is typically accomplished by high temperature calcination to remove coke residues from the porous structure of the zeolites.
- The tunability of the acid properties of the catalysts by modifying the Si/Al ratio allows precise control of the selectivity of the reaction by selecting the most appropriate acidity for each particular process.
- The possibility of selecting between a large number of different topologies with varied pore apertures and channel dimensionalities is doubtless the most unique feature that distinguishes zeolites from other acid solid catalysts. Indeed, zeolites can drive the selectivity of a process to the target

product by selecting the most appropriate topology. This property is generally called **shape selectivity** and is a direct consequence of the presence of structural—and therefore highly uniform—microporosity, with pore apertures in the range of most of the hydrocarbons of interest in refineries. Indeed, it is possible to select a particular zeolite, which allows the entrance of a particular reactant but not bulkier ones; this is generally called **reactant shape selectivity**. In some processes, the zeolite is selected because it allows the diffusion of only one of the different products formed on the acid sites, and therefore only that product can be recovered downstream—this is referred to as **product shape selectivity**. Finally zeolites with cavities on channel crossings can allow the formation of a reaction intermediate and not others, giving rise to the formation of a unique reaction product instead of a mixture of them—called **transition state shape selectivity**).

The presence of one or several of these properties improves the selectivity of the processes enormously and permits the engineering of continuous flow processes by the use of either single or multi-bed as well as fluidised reactors.

More than 170 zeolitic structures have been discovered with different topologies.[19] However, many of them can be obtained with a broad chemical composition range and thus there are a huge number of possible combinations of topologies and chemical compositions for catalyst optimisation.

But despite the large number of zeolites available, only a few have found commercial application in refineries, petrochemical processes or gas separations; examples include faujasite (FAU), ZSM-5 (MFI), Zeolite A (LTA), ferrierite (FER), mordenite (MOR) and SAPO-11 (AEL). There is no doubt that, Zeolite Y is the most important zeolitic catalyst, being the main component of the catalysts employed in fluidised catalytic cracking (FCC), with a total annual consumption over 550 000 tons/year.[20]

There are excellent reviews detailing the use of zeolites in refinery and petrochemical processes.[21–28] This chapter describes the most important catalytic applications of zeolites within the refinery scheme, under flow conditions, focussing on the benefits that have been obtained by the use of those microporous catalysts in the described processes. The application of zeolites in some key reactions in petrochemistry and fine chemical production are also outlined.

3.3.1.1 Zeolites in Refining Processes

Refinery processes have been developed to transform crude oil into valuable fuels, lubricants and commodities for petrochemistry. To afford these objectives, there are a large number of separation processes (adsorption, extractions and/or distillations) and chemical processes in which the relatively large molecules of the heavy oil are transformed into smaller ones with the properties appropriate to the final uses of the products. These processes (either catalytic or non-catalytic) are required because there is a large difference between the distribution and quality of the products obtained by direct distillation of the crude

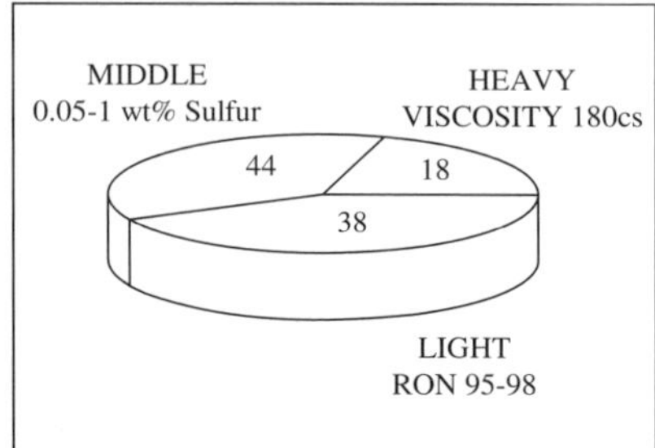

Figure 3.3   Differences between the direct crude-oil distillation and the market demand of petroleum derivatives. Sulfur content is given as weight percentage and viscosity units are anti-stokes. RON stands for research octane number.

oil and the market demands. This is illustrated in Figure 3.3, where the quantity and quality of the product distribution of light, middle and heavy distillates obtained from direct distillation of a Dubai crude oil are compared with the approximate actual requirements of the fuels and lubes (lubricants) needed for the market.

It is clear from Figure 3.3 that heavy distillates need to be converted to medium and light hydrocarbons but that the octane number, viscosity and sulfur content also need to be upgraded for the market. At this point, zeolites have marked important differences with respect to non-microporous acid catalysts such as silica–alumina or clays, permitting a dramatic increase in the selectivities of the processes towards the desired products.

It is clear from the simplified scheme of a refinery shown in Figure 3.4 that all the streams coming out of the crude oil distillation need further improvements; these are afforded through downstream treatments to give rise to the final products (fuels, lubricants, *etc.*). The practical totality of the products is obtained through a zeolite-based catalytic process in at least one of the steps of their production—these processes are briefly described below.

Isomerisation of Light Paraffins.   The 'light naphtha' stream is formed mostly by branched and lineal paraffins in the C_5–C_6 range. The isomerisation of this stream is performed to increase the branched isomers, which possesses much higher octane numbers than linear paraffins and can therefore be introduced into the gasoline fraction without penalty in the final octane number in the fuel.[28]

Because isomerisation of short paraffins is a highly acidic demand reaction, the catalysts for carrying it out need to have very strong acidities; typically two kinds of catalysts are employed industrially—chlorinated aluminas and zeolites. The very strong acidities of the former allow their use as isomerisation

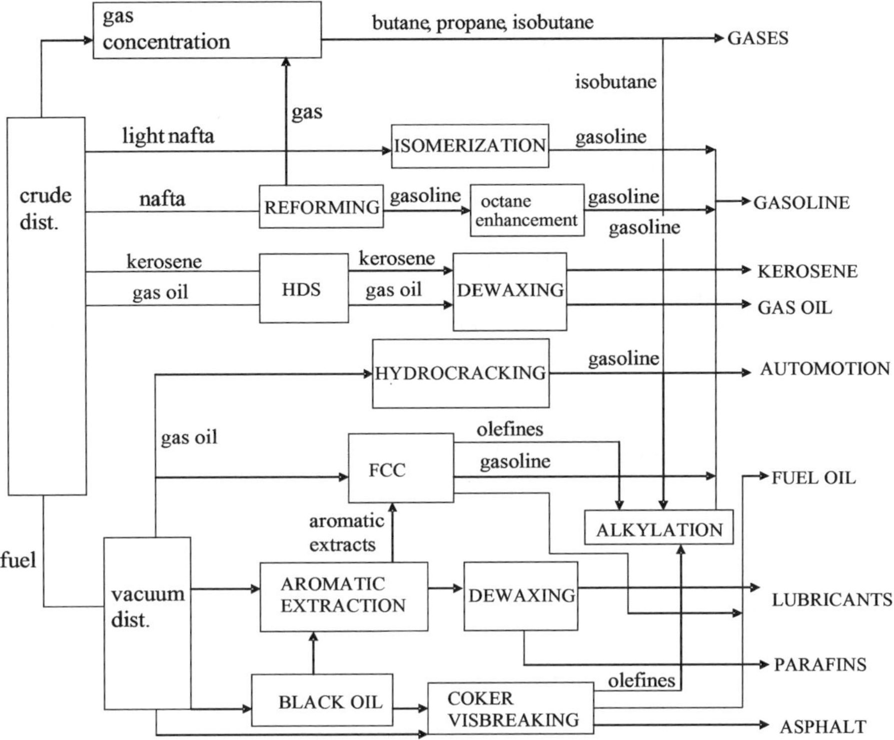

Figure 3.4 Diagram of production in a refinery.

catalysts at relatively low temperatures (180 °C), where the branched products are thermodynamically favoured. The major drawback of chlorinated aluminas is that:

- they are highly sensitive to water and sulfur;
- they need a constant addition of chlorine, since some HCl is constantly leached out of the catalyst causing serious problems of corrosion in the isomerisation plants where these catalysts are employed.

The second family of light paraffin isomerisation catalysts used in industry is mordenite-based catalysts.[29] Mordenite is perhaps one of the zeolites possessing the strongest acidity, though its acidity is still lower than that found in chlorinated aluminas. Consequently, the operating temperature of the isomerisation plants in which mordenite-based catalysts are employed is higher than that used when chlorinated aluminas are used as the isomerisation catalysts. Light paraffin isomerisation plants typically work at temperatures around 250 °C, which has a negative effect in the thermodynamic product distribution for obtaining branched paraffins. However, zeolite-based catalysts are less sensitive to water and sulfur, allowing their performance in the presence of relatively high levels of sulfur in the stream. This introduces an important benefit, because there is no need for deep desulfuration of the light paraffin stream prior to its isomerisation.

As it has been said above, the main zeolite used for this process is mordenite, which is de-aluminated to reach the most adequate Si/Al ratio (*i.e.* the acid strength) for this reaction. It is impregnated with a noble metal (generally platinum) in order to introduce a dehydrogenation/hydrogenation function in the final catalyst and avoid the formation of coke.

The most accepted mechanism for the isomerisation reaction of short paraffins relies on a bifunctional pathway. In the first step, the linear paraffins are dehydrogenated on the Pt catalyst and then the corresponding olefin is isomerised on the acid sites present in the zeolite structure. Finally, the resulting branched olefin is hydrogenated on the Pt sites introduced in the catalyst.[28]

The first process for isomerising the light paraffin stream was introduced by Shell in the 1960s with a process called Hysomer and based on the use of a Pt/ mordenite catalyst. Later on, UOP described a process, similar to the previous one, in which the isomerisation was integrated with the separation and recycling of the un-reacted linear paraffins (PENEX). In the 1990s, CEPSA and ITQ developed a new catalyst also based on Pt/mordenite (Hysopar), but the zeolitic component was modified to increase the resistance to sulfur poisoning.[29,30]

Sulfated zirconias have also been introduced as excellent candidates for isomerising light hydrocarbons. This is primarily based on their superacidic properties, which are capable of reducing the temperature of reaction significantly; thus, the thermodynamic equilibrium is shifted to a higher production of the desired branched hydrocarbons.[31,32] However, the instability of the sulfur phase against reduction and loss as hydrogen sulfide (H_2S) must be solved before a final industrial application can be found.

Catalytic Reforming of Light Gasoline.   Methyl-tertiary butyl ether (MTBE) has played a key role in the development of low contaminant fuels. However, the discovery of MTBE in surface water and groundwater has prompted a review of the presence of MTBE as a blender in gasolines. In addition, the strict regulation of aromatics, olefins and sulfur in fuels has prompted refineries to reformulate automotive combustibles to maintain current octane numbers using more environmentally friendly technologies based on high-octane number hydrocarbons instead of the MTBE technology. This could come from the use of highly branched paraffins with 7–10 carbon atoms obtained by isomerisation of linear paraffins. Therefore, the production of a paraffin-based high-octane gasoline blend stock (*e.g.* isomerates from isomerisation of light and mid-cut naphtha) could be a key technology for fuel supply to cope with future gasoline regulations.

Nowadays, the heavy naphtha stream (C_7–C_9) is subjected to an aromatisation process (catalytic reforming) to increase its octane number by creating aromatics (with a higher octane number than paraffins) from linear paraffins and naphthenes, and also through the hydroisomerisation of paraffins. Even though currently there is a strong limitation in the aromatic content in fuels, this process is still important because of the high added value of aromatics for petrochemistry.[20b]

Finally, the octane number of the resulting gasoline could be improved by hydrocracking the residual low-octane linear paraffins of this stream. Zeolites have been used to perform the hydrocracking step by taking advantage of the ability of some zeolites to allow diffusion of linear paraffins but not of the branched ones; only those which can diffuse inside the zeolite pores and reach the active acid sites will suffer the cracking reaction. Following this idea, Mobil introduced in the 1960s the Selectoforming process based on Erionite,[33] which was later modified by the M-forming process based on ZSM-5 with a better selectivity towards branched paraffins.[34] In addition, ferrierite has been employed for the catalytic reforming of naphtha in a process developed by Gulf. However, there is a limit to this process because the hydrocracking reaction reduces the yield of gasoline accompanying the improvement of the octane number, giving way to an increase in low value saturated gases.

Direct hydroisomerisation of linear paraffins has been attempted using Pt/mordenite but, unfortunately, the strong acidity of these zeolites produces a severe cracking reaction of paraffins higher than C_7 and cannot be used in this process.

Other less acidic zeolites have been proposed as candidates for this process; large pore zeolites such as Beta and USY zeolites, or medium pore zeolites such as Al-ZSM-5 or B-ZSM-5, with well-tailored acid strength have been proposed as appropriate candidates for the hydroisomerisation of heavy naphtha.[35–39]

In particular, mild acid catalysts based on Pt-supported on SAPO molecular sieves have been used. High yields of methylhexanes and dimethylpentanes were observed during isomerisation of heptane on Pd/SAPO-11 and Pd/SAPO-5, with a very low cracking rate, as could be expected from their mild acidity. Also, Pt/SAPO-11 was able to convert *n*-octane by hydroisomerisation to monobranched C_8 paraffins, while the cracking reaction took place preferentially on larger pore size Pt/SAPO-5. The unique activity and selectivity towards monobranched paraffins of Pt or Pd/SAPO-11 increased the higher the alkane, $n\text{-}C_6 < n\text{-}C_8 < n\text{-}C_{16}$.[40–43] Unfortunately, no SAPO catalyst has been found able to give that exceptional selectivity towards multibranched paraffins, which have a much higher octane number than monobranched hydrocarbons. However, this catalyst can find application in isodewaxing processes as described below.

Zeolites have also been applied to the reforming of the C_6 and C_7 fractions to give rise to toluene and xylenes. In 1996, EXXON-Mobil introduced the BTXtra[SM] process based on a zeolite (details have not been disclosed) which is able to increase the yield of toluene and xylenes. This zeolite could *trans*-alkylate with benzene and/or de-alkylate trimethylbenzenes to form benzene and mostly toluene and xylene. Given the high value of the resulting products, this process could be said to be between refining and petrochemistry. In this regard, the best selectivity for the aromatisation of these fractions has been found with a Pt-supported catalyst on a Ba and K exchanged-Zeolite L.[44]

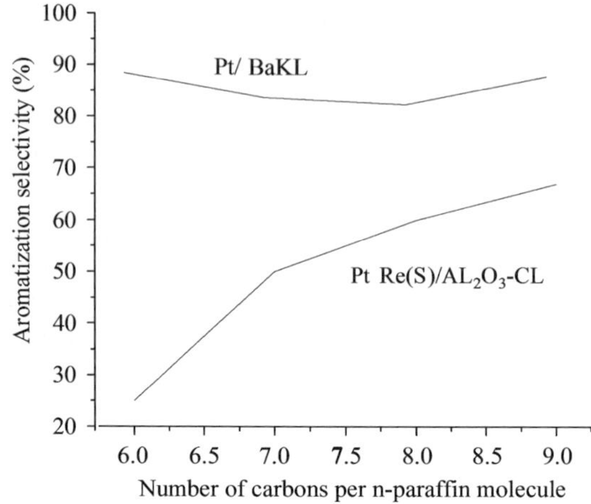

Figure 3.5   Catalytic reforming with zeolite L.

This catalyst has shown a much better performance for catalytic reforming of C_6, C_7 or even C_8 and C_9 fractions as illustrated in Figure 3.5, where its selectivity is compared to that of the commercial catalyst.[20a]

The main drawback of the Pt/Ba, K-Zeolite L catalyst is its extreme sensitivity towards the presence of sulfur in the feed. This is a clear consequence of the low acidity of the exchanged zeolites, which strongly reduces their thioresistance. Chevron and UOP have commercialised processes based on this technology called AROMAX and RZ100, respectively.

Dewaxing and Isodewaxing of Middle Distillates.   The dewaxing processes are of interest in order to improve the performance of middle distillates (kerosene and diesel) and lubricants. This is because linear paraffins have higher melting points and they crystallise at relatively low temperatures, giving rise to problems of filter blocking, high viscosities, or even fuel freezing in extreme low temperature conditions such as those found for jet fuels.

Dewaxing can be afforded through different technologies such as solvent extraction, selective cracking of paraffins and isomerisation of linear paraffins. In the case of middle distillates, only cracking and isomerisation are commercially employed.

ZSM-5 zeolite is the most widely used catalyst for dewaxing since it was introduced by Mobil in 1978 (MDDW process).[45] This catalyst produces a selective cracking of the paraffins as only they are small enough to penetrate inside the ZSM-5 pores. This process was further improved by introducing a second desulfuration function in the catalyst, which allows the removal of the sulfur as well as the saturation of olefins and aromatics (CFI process). More recently, a combined MDDW–CFI process developed jointly by Mobil, Akzo and Kellog improved the engineering, allowing the production of high quality kerosene.

However, the cracking reaction of middle distillates could be an undesirable option for refineries in countries with a high demand of diesel for automobiles, when isomerisation of this fraction could be a better option. ZSM-5 and SAPO-11 are the two catalysts used to do this commercially at present. ZSM-5 was introduced by Mobil in 1990 (MIDW process) to produce diesel with a very low pour point and a low level of sulfur.

Pt or Pd supported on SAPO-11 produce a selective isomerisation of linear paraffins to monomethyl saturated hydrocarbons (as was seen above). These monobranched hydrocarbons reduce the pour point and viscosity of this fraction, and also maintain a very good cetene number in the diesel fraction.[46]

Fluid Catalytic Cracking of Vacuum Gas Oil. As shown in Figure 3.4, the bottoms of the atmospheric distillation are sent to be treated in a second vacuum distillation tower; a vacuum gas oil is obtained at the head of this unit. Other intermediate distillates of this vacuum distillation are raw materials for lubricants, and the bottoms of the vacuum distillation can be revamped into lubricants by de-asphaltation with propane. However, the most important fraction is the vacuum gas oil, which can be converted into valuable fuels and olefins by cracking or hydrocracking processes. The catalytic cracking of vacuum gas oil is carried out in the fluid catalytic cracking (FCC) unit. Here the cracking reaction takes place in a mobilised bed and is currently the main process in which zeolites are utilised for catalytic uses.

In the FCC unit, the gas oil is mixed with the catalyst at the entrance of the reactor (called the raiser). They rise together at very high speed (the contact time is of the order of a few seconds) through the reactor, giving rise to the cracking reactions until the separation unit or stripper is reached. In this part of the plant, the catalyst and the products are split and the catalyst, which is deactivated by coke deposits, is regenerated by air calcination of the coke, being recycled back into the raiser.

The FCC products are mainly made up of:

- gases—mostly light olefins such as ethylene, propylene and butylenes of high value for petrochemistry;
- high octane gasoline with a high content of olefins and aromatics, which needs to be improved;
- LCO (light cycle oil)—is a stream of heavy diesel with a high concentration on aromatics, which can be used directly for automobiles, heating or as bunker (ship fuel);
- unconverted heavy fractions, which can be recycled to the FCC or any other thermal cracking process;
- coke that is burnt in the regeneration unit and which serves to maintain the energy balance of the FCC process.

The FCC catalyst is based on Zeolite Y—a tridirectional large pore zeolite with the faujasite structure. These zeolites are typically synthesised with low Si/

Al ratio (*i.e.* with very high Al content) and thus possess a high concentration of acid sites in the medium range of acid strength. The presence of a high concentration of Al in the zeolites also reduces the thermal and hydrothermal stability of the catalyst, which becomes unstable under the regeneration conditions (temperatures close to 700 °C in the presence of steam). Therefore, catalyst stabilisation is required.

The first generation zeolitic catalyst employed in FCC units was based on rare earth exchanged Zeolite Y (REY) because the presence of exchanged cations in the faujasite structure reduces the Brönsted acidity of the zeolites, thus increasing the thermal and hydrothermal stability of the catalyst. However, due to the reduction of the acidity of the zeolites, the REY catalysts were not more much active than amorphous silica–alumina, though they provided a better selectivity towards low octane gasoline and middle distillates.[47] The reason for the low octane of the gasoline produced on REY catalysts is the high concentration of acid sites (even if these are not very strong), favouring the hydrogen transference reactions between olefins and naphthenes. Therefore, the selectivity is shifted to aromatics and paraffins, which leads to low octane numbers (as seen above).

In 1964, Grace-Davison introduced Ultrastable Zeolite Y (USY), which was produced by deep de-alumination of proton-exchanged zeolites Y (HY) by steam calcination at high temperature or through acid treatments. The advantage of USY catalysts is that they possess a lower concentration of acid sites, but of higher strength than previous REY catalysts. Hydrogen transfer reactions are limited and high-octane gasolines are produced. In addition, USY catalysts increase the yield of gasoline and reduce coke formation—even though their intrinsic activity is slightly lower than that of REY catalysts. The first USY catalyst was commercially assayed in 1975, and since then the practical worldwide totality of FCC catalysts is based on USY zeolites.

Subsequently, a new family of catalysts which combines the benefits of REY and USY zeolites was commercialised. These catalysts are based on rare earth exchanged USY zeolites (REUSY) and provide a better motor octane number (MON) than conventional USY (Table 3.1).[48]

Table 3.1 Composition and octane number of FCC gasoline obtained using different Zeolite Y-based catalysts.

Gasoline composition	Catalysts		
	REY	*USY*	*REUSY*
Paraffins: linear	2.1	2.0	2.3
branched	21.5	14.7	16.4
Olefins: linear	12.1	11.6	13.1
branched	19.4	23.1	23.6
Aromatics	26.8	24.8	27.1
RON (research octane number)	89	94	94
MON (motor octane number)	78	79	82

In 1986, Mobil introduced the use of ZSM-5 (a medium pore zeolite) as an additive for the FCC catalyst to increase the yield of light olefins (high importance in petrochemistry) and the octane number of the gasoline.[48,49] The effect of the ZSM-5 additive is similar to that described for catalytic dewaxing with low-octane linear paraffins being cracked to light olefins (propylene and butylenes).[20] In this sense, REY catalysts, which produce the highest yield of linear paraffins, are the most sensitive to the presence of ZSM-5 in the FCC catalyst. The increase in the octane number is accompanied by a reduction in gasoline yield; this is becomes more important as the amount of ZSM-5 additive in the FCC catalyst formulation increases (Figure 3.6).

In addition to USY and ZSM-5 zeolites, the FCC catalyst possesses a large number of other components including:

- active matrixes responsible for precracking of the bulkiest molecules (very useful for cracking residues);
- clays that introduce sodium resistance;
- a binder, which introduces the mechanical strength required to be fluidised;
- traps of metals such as V and Ni.

The FCC is probably the most versatile process within the refinery and can be conveniently modified depending on the characteristics of the feed. At the same time, the selectivity of the process can be changed by controlling the operating conditions or modifying the catalyst, which is continuously introduced into the unit. Thus, olefin production can be maximised for petrochemicals or for alkylation gasoline (isobutene + butene for C_8) or MTBE (methanol + isobutene), or it can be operated to maximise the yield of gasoline or LCO depending on the seasonal demand.

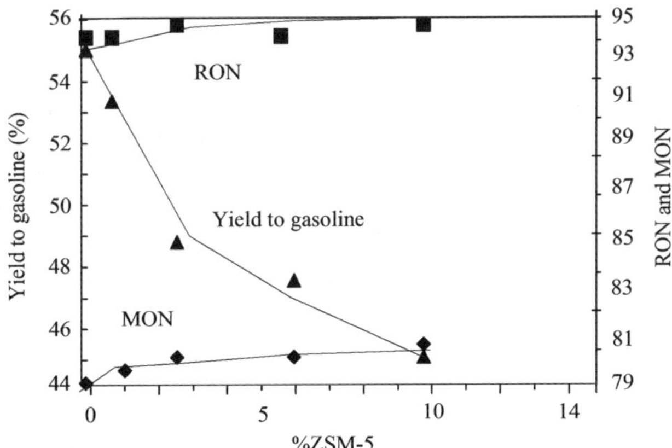

Figure 3.6　Effect of the amount of ZSM-5 additive in the FCC catalyst. Triangles, squares and diamonds stand for yield of gasoline, research octane number (RON) and motor octane number (MON) respectively.

Hydrocracking. Hydrocracking is a refinery process in which heavy fractions are converted into lighter ones but, at the same time, the H/C ratio increases with respect to the original feed without producing coke. This reaction takes place at high temperatures (350–450 °C) and hydrogen pressures (100–180 bar), making it one of the most expensive units in refineries. In general, this reaction is carried out using bifunctional catalysts having an acidic function, such as those present in silica–aluminas or zeolites, and a second hydrogenating/dehydrogenating function provided by the presence of a Nobel metal (Pt, Pd) or by a combination of sulfides of Mo and W, promoted by Co and/or Ni.[50,51] The acid function is typically supplied by USY zeolites, since they are highly acidic and allow operation at lower temperatures (being more resistant to poisoning by sulfur and nitrogen). The level of de-alumination of the USY catalyst permits control of the selectivity of the process towards middle distillates, increasing as the Al content in the catalyst falls.[51]

There is a process—called mild hydrocracking (MHC) process—which converts vacuum gas oil into middle distillates and fuel with very low sulfur content at much lower pressures (30–70 bar) and temperatures (350–440 °C) than the conventional hydrocracking process. The MHC process is catalysed by Mo/Co supported on USY zeolites.

Dewaxing and Hydrodewaxing of Lubes. The heavier products obtained in the vacuum distillation tower after the vacuum gas oil are the lubes, which need to be treated with furfural to extract the aromatics. These refinates need to be dewaxed to increase their cold properties (pour point) and to undergo a process called hydrofinishing to remove sulfur and improve their colour.

The dewaxing of lubes is typically carried out by a catalytic process. This process can consist of either the cracking or isomerisation of paraffins. In both processes, zeolites are used as catalysts since their shape selectivity allows selective treatment of linear paraffins as discussed above.[52] The first such zeolite-based process was introduced by BP in 1972 and was based on the use of Pt/mordenite, though the relatively large pore aperture of this zeolite (a unidirectional large pore zeolite) produces the cracking of isoparaffins with a negative effect on the viscosity. In the 1980s, Mobil introduced a modified ZSM-5, which is more widely used. It has better selectivity towards cracking of linear paraffins, but still presents problems of low yield and viscosities when compared with lubes obtained by chemical dewaxing.[53]

Chevron introduced a new process based on the isomerisation of linear paraffins using SAPO-11 as a selective catalyst.[54] This process does not reduce the yield of lubricant and maintains the viscosity index. The reasons for this high selectivity are the unidirectional medium pore channel system and the very mild acidity present in SAPO-11, which promote the isomerisation of paraffins *versus* the cracking reaction.[55]

The benefit of using isodewaxing processes instead of chemical extraction of lubricants is that:

- the former requires lower capital inversion and has a lower operating cost than the latter;
- the products have a better pour point.

3.3.1.2 Zeolites in Petrochemistry

The utilisation of zeolites in petrochemical processes derives from the production of aromatics and their ability to provide extremely convenient shape selectivity in the product distribution.[56]

The main source of aromatics for petrochemistry originates from the catalytic reformate of naphthas. The stream coming out of the catalytic reformer is subjected to a chemical extraction of the BTEX aromatics (benzene, toluene, ethylbenzene and xylenes) before being separated by distillation and interconverted by catalytic transformation, very often using zeolites as catalysts. Figure 3.7 provides a basic schematic of a petrochemical plant.

First, benzene and toluene are separated by distillation from the ethylbenzene and xylenes mixture. Then, the *ortho*-xylene is distilled from this complex mixture and the remaining mixture is processed to obtain *p*-xylene through the following three processes.

Xylene Isomerisation. The objective of this process is to maximise the production of *p*-xylene until the equilibrium condition is reached. Frequently, this implies the need to treat the xylene mixture with the ethylbenzene present in the reformate stream since ethylbenzene is very difficult to distil out.

Mobil has commercialised two processes based on ZSM-5. The first one isomerises the xylene and converts the ethylbenzene in benzene and diethylbenzene, which can be now removed *via* a *trans*-alkylation reaction. The pore diameter of the ZSM-5 zeolite allows the rapid diffusion of *p*-xylene, while *o*-xylene, which diffuses much slower, isomerises until the *para*-isomer is produced (shape selectivity towards the product). In addition, the topology of ZSM-5 zeolite avoids the bimolecular reaction of xylene disproportionation because the intermediate state cannot be formed inside its pores (transition state shape selectivity).[56b,57,58] The combination of these two effects has allowed the commercial use of ZSM-5 as a catalyst for *p*-xylene production. In the second process commercialised by Mobil, the ethylbenzene is dealkylated to benzene and ethane on Pt/ZSM-5 catalysts.

Toluene Disproportionation. The objective of this process is to convert toluene into *p*-xylene. Toray has commercialised a process based on mordenite, which allows the conversion of toluene into the thermodynamic mixture of xylenes as well as the transalkylation of toluene with trimethylbenzenes or

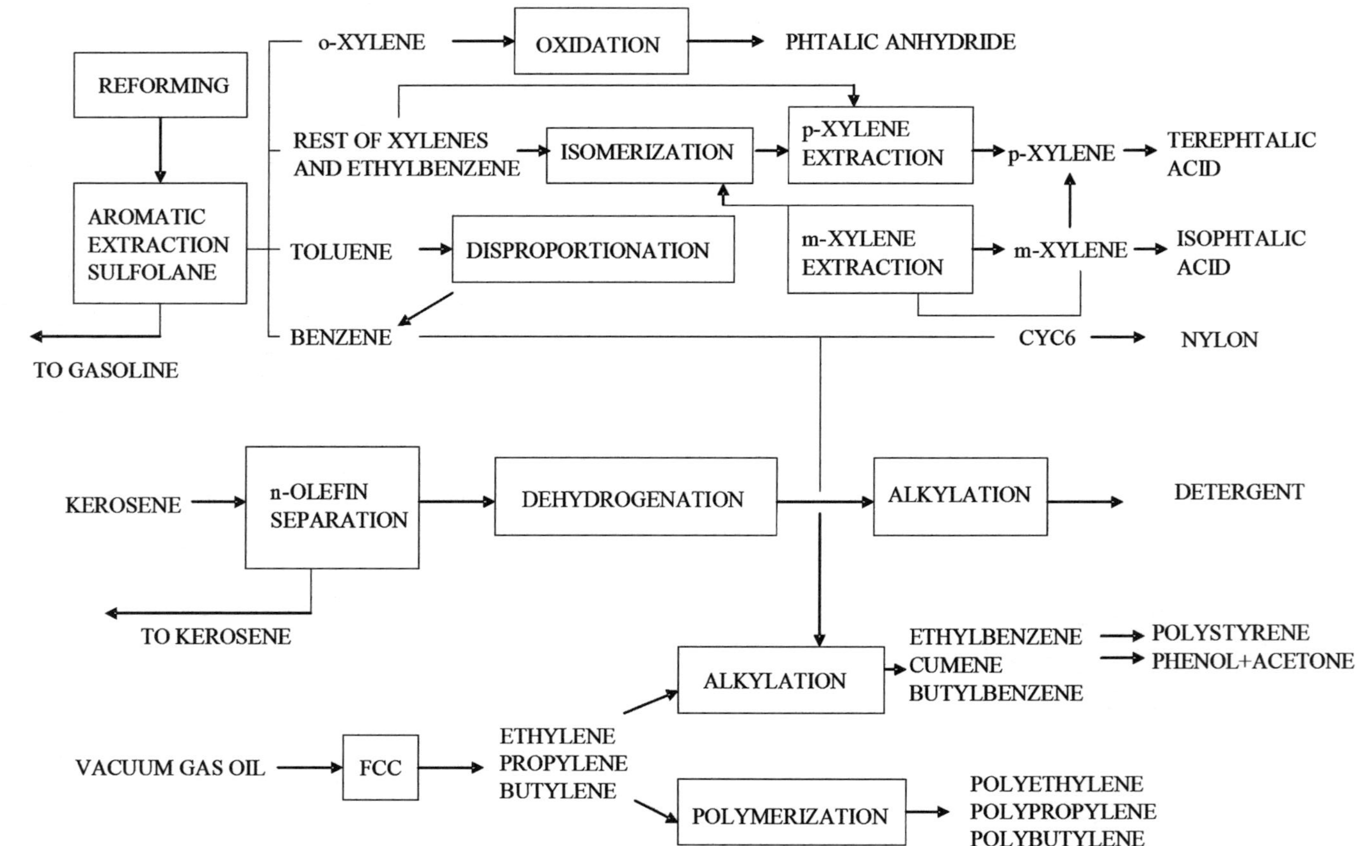

Figure 3.7 Diagram of a petrochemical plant (processes catalysed by zeolites are shown in boxes).

higher alkylaromatics to form xylenes. However, the processes based on ZSM-5 maximised the selectivity towards *p*-xylene. To obtain a maximum selectivity, ZSM-5 is 'selectivisated' by coking or silica deposition to narrow the pore apertures and thus make the diffusion of *o*-xylenes difficult.[59,60] Using 'selectivisated' ZSM-5 catalysts, which combine acid properties with the appropriate diffusion characteristics, it is possible to obtain up 80% selectivity towards *p*-xylene (by toluene disproportionation).[20c,56b]

Synthesis of Ethylbenzene and Cumene. Nearly 90% of the ethylbenzene produced nowadays is consumed in the manufacture of styrene–polystyrene. It is produced by direct alkylation of benzene with ethylene, a reaction typically performed on acid catalysts such as $AlCl_3$/alumina, BF_3/alumina or zeolites. The Mobil process uses ZSM-5 as catalyst, which possesses an adequate porosity to carry out this reaction. However, large quantities of diethylbenzene are also produced during the course of the process. The diethylbenzene is transformed into the desired ethylbenzene *via* a *trans*-alkylation process with benzene. The overall yield of the process is 99% to the desired monoalkylated benzene.[56b] Recently, a new zeolite, MCM-22, has been commercialised for this process. Its high intrinsic activity allows this reaction to be carried out in liquid phase.[20c]

The alkylation of propylene on ZSM-5 zeolite yields the formation of mostly normal propylbenzene instead of the desired cumene (2-propyl-benzene). For this process, other zeolites have been applied with larger pore apertures such as mordenite, Beta and faujasite.[56b]

3.3.2 Current Applications in the Fine Chemicals Industry

Environmental reasons mean that the fine chemicals industry must adapt to more and more strict regulatory constraints in a short term. In this respect, several approaches enable the improvement of chemical processes in ecological terms.[61a] One of them consists of applying the concept of specific processes functioning continuously through the use of selective and robust catalysts such as zeolites.[61b] This concept offer new opportunities for the fine chemicals industry such as:

(a) developing new reactions that minimise discharges and waste by carrying out reactions in the vapour phase with a fixed bed catalyst;
(b) the option to modify the selectivity with regard to reactants, intermediates and products;
(c) simplification of processes (*i.e.* simpler separation and purification of reaction products).

This section describes examples of flow systems in the production of fine chemicals, showing that zeolites can be successfully applied to replace environmentally undesirable methodologies.

3.3.2.1   *Synthesis of ε-caprolactam*

Nylon-6 is produced from the ring-opening polymerisation of ε-caprolactam, which is commercially produced by a liquid phase Beckmann rearrangement of cyclohexanone oxime in highly concentrated sulfuric acid (oleum) and a combination of processes with a high environmental impact (route A, Scheme 3.1).

An alternative and cleaner route to ε-caprolactam has been devised starting from a selective partial hydrogenation of benzene to cyclohexene, followed by a hydration reaction of the latter in the presence of H-ZSM5 to afford cyclohexanol (route B, Scheme 3.1). Asahi Kasei commercialised this hydration step using high-silica HZSM-5 zeolite (although in batch reactor) in 1990.[62] Since then, the process has been operated successfully on the scale of 60 000 tons/year.[62]

After oxidation of cyclohexanol to cyclohexanone, the oximation to cyclohexanone oxime is carried out in the presence of ammonia and hydrogen peroxide as the oxidant over a Ti-containing MFI-zeolite TS-1 (Enichem Process, 12 000 tons/year demonstration plant).[63] Zeolite TS-1 gives excellent yields of cyclohexanone oxime at moderate temperatures. Thus, for example, at a reaction temperature of 60 °C, cyclohexanone conversions above 90% and selectivities above 99% for the oxime have been reported. Finally, the oxime rearranges to ε-caprolactam through a Beckmann rearrangement.

While many routes to the oxime exist, all commercial caprolactam production makes use of a Beckmann rearrangement to transform the oxime to the desired product. So far, the heterogeneously catalysed rearrangement last step has not been commercialised because of problems such as catalyst lifetime. These drawbacks still have to be solved as it would not make sense to manufacture the oxime by the use of a heterogeneous catalyst and still carry out the Beckmann rearrangement in a homogeneously catalysed procedure.

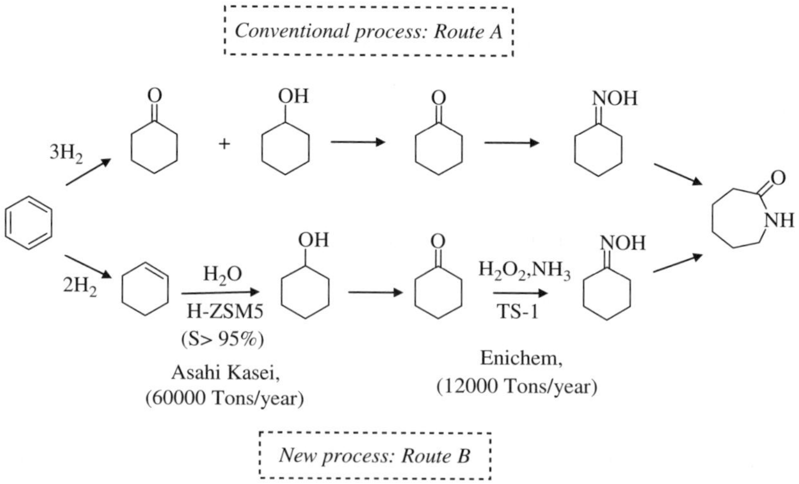

Scheme 3.1   Synthetic sequences from benzene to ε-caprolactam.

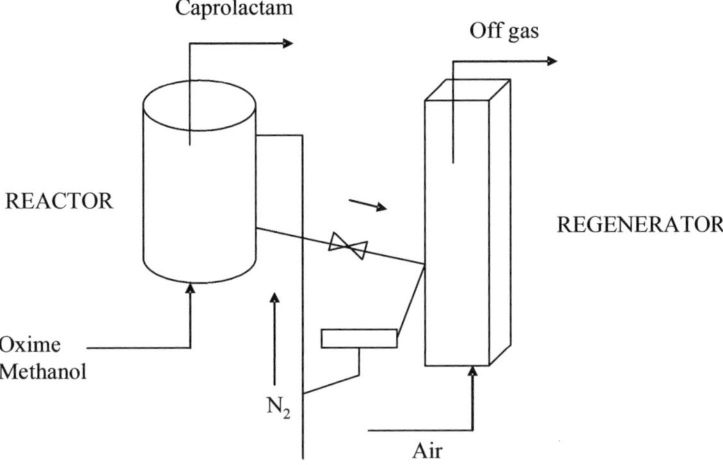

Figure 3.8 Fluidised bed reaction system.

In close connection to this, research has focused on the gas-phase rearrangement of cyclohexanone oxime using solid acids.[64] In the use of zeolites as catalysts, Ichihashi and co-workers at Sumitomo developed a catalytic vapour phase Beckmann rearrangement over a high-silica MFI zeolite.[64] The process adopts a fluidised bed reaction system.[65] As shown in Figure 3.8, the mixture of cyclohexanone oxime and methanol vapour is fed into a fluidised bed reactor in which the catalyst is charged.[64c,65]

Cyclohexanone oxime is converted to caprolactam in good yield and high efficiency. Part of the catalyst is transferred from the reactor to the regenerator and is reactivated; the regenerated catalyst then returns from the regenerator to the reactor. The catalyst circulates continuously through the reactor and the regenerator.

3.3.2.2 Synthesis of Trioxane

Gas hydrations of different olefins have also been examined using various zeolites in which pentasil and ferrierite zeolites showed high activities.[66] This example features trioxane, a raw material for polyacetal copolymer, which is synthesised from formalin (aqueous formaldehyde) using sulfuric acid. Because the equilibrium concentration is low (4–5%), isolation of trioxane is carried out by distillation at 373–393K. However, under these conditions, the resulting formic acid (by-product) solution causes corrosion of the equipment. Attempts by Asahi Kasei to use zeolites for the trioxane synthesis are presented in the Scheme 3.2.[67]

The synthesis was carried out in a glass tube reactor equipped with a water condenser, to which formalin (65% w/w) and zeolite were introduced at 373K. In this case, zeolites with Si/Al >8 were active. Among the high-silica zeolites,

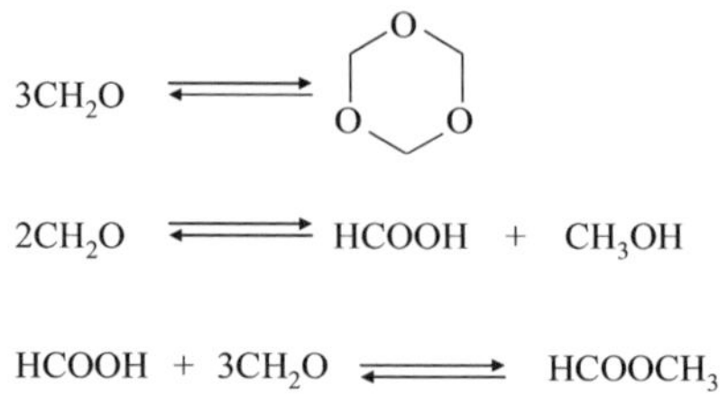

$$3CH_2O \rightleftharpoons$$

$$2CH_2O \rightleftharpoons HCOOH + CH_3OH$$

$$HCOOH + 3CH_2O \rightleftharpoons HCOOCH_3$$

Scheme 3.2 Reaction paths of formaldehyde in water.

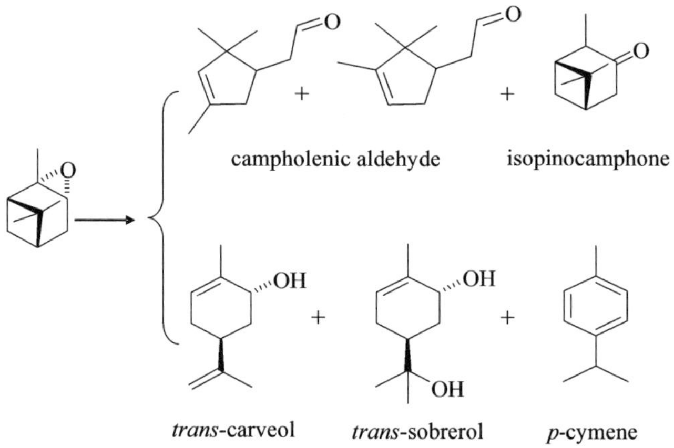

campholenic aldehyde   isopinocamphone

trans-carveol   *trans*-sobrerol   *p*-cymene

Scheme 3.3 Products obtained By α-pinene oxide rearrangement.

β zeolite (Si/Al ratio = 14) was $\sim$10 times more active than Amberlyst-15 on a unit per catalyst weight basis. Therefore, similarly as above, these results also indicate that hydrophobicity plays an important role in controlling the catalytic activity of this reaction.

Treatment of HZSM-5 with 4-methylquinoline, which is adsorbed only on the external surface, did not result in any change in catalytic activity.[68] In contrast, treatment with pyridine, which is adsorbed also in the inner surface of the pores, effectively decreased the activity. These results indicate that the reaction takes place within the pores.

3.3.2.3 *Isomerisation of α-Pinene Oxide*

α-Pinene oxide is a very reactive substrate which rearranges under acidic conditions yielding a variety of products as shown in Scheme 3.3.[69]

Among these compounds, the industrially most desired one is campholenic aldehyde, an important intermediate for the synthesis of sandalwood-like fragrance chemicals such as santalol.[69,70] In a catalyst screening, numerous heterogeneous systems such as transition metal oxides, phosphates and zeolites have been checked in continuous as well as in slurry reactors at various temperatures, residence times, solvents, *etc.*[71] From these studies, it could be deduced that the tridimensional, large pore system of USY zeolite was a very suitable catalyst for this reaction.

Operating at near 0 °C with an acid-washed catalyst, the yields were not far from 85%—a value previously afforded by zinc halides. The good performance of USY was attributed to the presence of highly dispersed Lewis acidic Al centres.

The same reaction was catalysed by Ti-β zeolite both in liquid and gas phase conditions. In this case, initial yields up to 93% were achieved at 90 °C when co-feeding a gaseous stream of α-pinene oxide and the inert organic compound (*e.g. n*-heptane, methylcyclohexane) into a continuous-flow fixed bed reactor.[72]

The function of the organic diluent was to compete for adsorption on Ti-β, reducing the intraporous concentration of α-pinene oxide and accordingly its bimolecular side reactions. This, together with shape selectivity effects, was invoked to explain the high selectivity in a reaction involving a rather sensitive substrate and significantly higher temperatures than with other Lewis acid catalysts.

It is noteworthy that operation at these industrially practicable temperatures appears possible with Ti-β while, for other catalysts, low temperatures are required for high selectivity. Complete regeneration of the catalytic activity (up to 100 times) could be achieved by an air burn-off at 480 °C.

3.3.2.4   *Synthesis of p-Methoxyacetophenone*

One of the first industrial applications of solid acids in the acylation field introduced by Rhône-Poulenc (now Rhodia) was the liquid phase acylation of anisole with acetic anhydride over HBeta zeolite in a recyclable fixed bed reactor (Scheme 3.4).[73]

One of the important drawbacks of the conventional acylation process is the significant amount of waste in the form of inorganic salts such as $Al(OH)_3$ or NaCl, which are formed in subsequent neutralisation steps. Against this, the catalytic process developed by Rhodia has more advantages from the environmental point of view. The salient features of both processes are detailed in Figure 3.9.

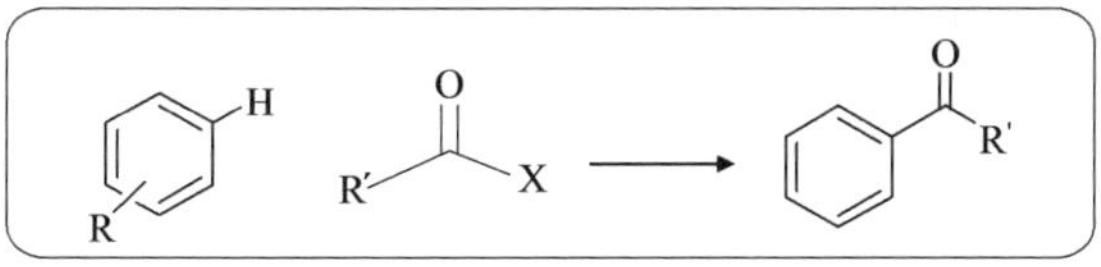

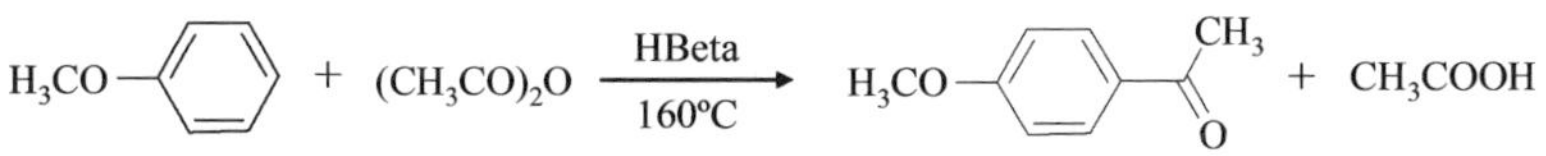

Scheme 3.4   Acylation of anisole with acetic anhydride over HBeta zeolite.

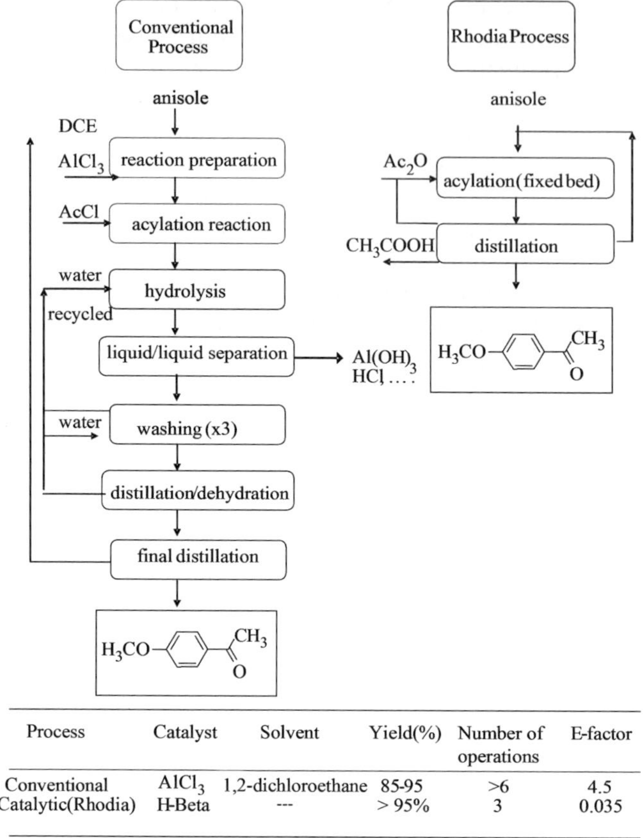

Process	Catalyst	Solvent	Yield(%)	Number of operations	E-factor
Conventional	AlCl₃	1,2-dichloroethane	85-95	>6	4.5
Catalytic(Rhodia)	H-Beta	---	> 95%	3	0.035

Figure 3.9   Comparative routes for the synthesis of *p*-methoxyacetophenone.

The replacement of the Lewis acid AlCl₃ by the acidic HBeta zeolite, with the considerable reduction in the number of operations as well as the important reduction in the E-factor (number of kg of waste produced per kg of product[61a]) are the most salient features of this new catalytic industrial process.

However, besides the shape selectivity and acid strength provided by acidic zeolites, sorption properties can emerge in some cases as a key factor in determining the course of the reaction. In this respect, it is necessary to indicate that, for the acylation of apolar aromatic compounds, the preferential sorption of the carboxylic acid and the retention of the carbonyl product actually led to unbalanced reagent ratios in the proximity of acidic sites and to inhibition, respectively, with negative effects on the rate.[69b]

On the basis of these results, there have been numerous studies ranging from mechanistic aspects to process optimisation. Solid acids considered include sulfonic resins, amorphous oxides, clays and zeolites; the range of acetylating agents tested cover acetic acid and its esters, acetyl chloride and acetic anhydride. A detailed study carried out by Rohan *et al.* revealed the function of reversible and irreversible adsorption phenomena in the decay of the catalytic activity.[74] Some of the results highlighted the advantages obtainable under flow conditions. The retention of *p*-methoxyacetophenone, assisted by its larger molecular size and relatively high polarity, caused product inhibition. This could be minimised by continuous-flow mode operation, using anisole-rich mixtures and solvent washing of exhausted catalyst. Activity tests in a fixed bed reactor, with anisole playing the dual role of reagent and solvent showed that β-zeolite remained fairly stable for *ca.* 50 hours before deactivation.[74–78]

3.3.2.5   *Industrial Route to 4-Methylthiazole*

Other reactions such as the dehydration of alcohols, the isomerisation of olefins (isomerisation of 1-butene), cyclisations (cyclo-addition of carbon dioxide to ethylene oxide) and side-chain alkylation of aromatics (alkylation of toluene with methanol or ethylene) are important reactions catalysed under basic conditions. Many of these processes are carried out industrially using liquid bases as catalysts. These applications may require nearly stoichiometric amounts of the liquid base for conversion to the desired product. Therefore, they result in strong pollution of the environment and open a wide and important research field to develop energetically and economically favourable as well as environmentally friendly alternative routes through the use of solid base catalysts. Typical solid basic catalysts such as X zeolites (faujasite) modified by cation exchange, impregnation, or isomorphous substitution are a clear alternative.

One example of an industrial process that utilises a basic zeolite catalyst, which has not been commercialised so far but has reached the pilot plant scale, is the process developed by Merck and Co. for the synthesis of 4-methylthiazole. In this case, Merck used caesium sulfate impregnated ZSM-5 as catalyst for the synthesis of 4-methylthiazol (Scheme 3.5).[79,80]

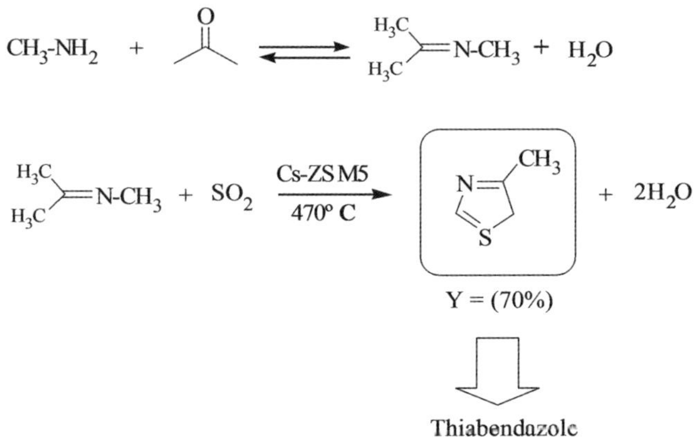

Scheme 3.5   New Merck route to 4-methylthiazole.

4-Methylthiazole is an intermediate in the synthesis of thiabendazole, which is used as a systemic fungicide. The current industrial route involves numerous reaction steps that use hazardous chemicals. Workers at Merck[81] reported that caesium-loaded ZSM-5 gives excellent activity, selectivity and lifetime for the synthesis of this compound from the imine and sulfur dioxide (SO_2). The reaction was run continuously for more than 170 hours. The catalyst showed an overall conversion of 86% and a high selectivity to 4-methylthiazole based on the gas phase imine content.

There are three important features in this work:

(a) A zeolite catalyst is used for the production of a pharmaceutical.
(b) The zeolite performs a basic catalysis.
(c) The new process is environmentally sound.

3.3.2.6 New Developments in the Production of Methyl Methacrylate

Methyl methacrylate is an important monomer which is widely used for producing acrylic plastics (polymethyl methacrylate) or producing polymer dispersions for paints and coatings. Methacrylic polymer, which has the characteristics of good transparency and weather resistance, is used in many fields such as signboards, building materials, vehicles and lighting equipment.

Methyl methacrylate can be produced in different ways from C_2–C_4 hydrocarbon feedstocks,[81] although the conventional method is based on a C_3 route. Mitsubishi Gas Chemicals has industrialised several improvements to this route. In the new process, the use of sulfuric acid is avoided, a *trans*-esterification reaction is performed without methanol but with methylformate, and finally the last step (dehydration of methyl α-hydroxyisobutyrate) is carried out on a zeolite catalyst in the gas phase (Scheme 3.6).

Initial problems for this procedure such as the presence of a side reaction or the reduced catalyst lifetime were solved using methanol and a faujasite zeolite.[82] In this case, the process consisted of continuous feeding of a mixture of methanol and α-hydroxyisobutyrate in a 0.1–3.0 ratio (w/w) to the reactor and carrying out the reaction at high reaction temperature using a NaCsY zeolite as catalyst. The process is capable of producing the target product with high yield

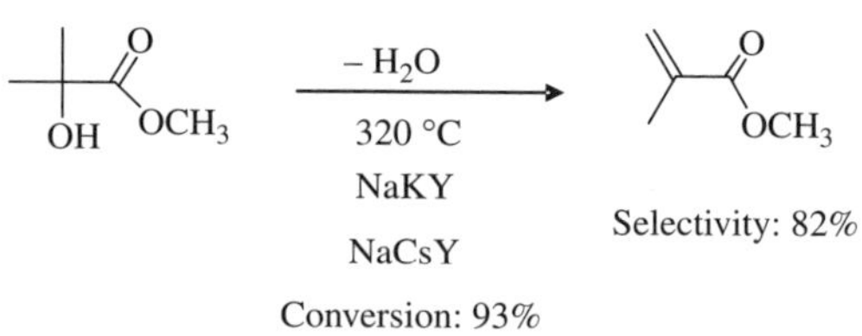

O
OH OCH₃
– H₂O
320 °C
NaKY
NaCsY
Conversion: 93%
O
OCH₃
Selectivity: 82%

Scheme 3.6 Dehydration process of methyl α-hydroxyisobutyrate in the presence of zeolites in the gas phase.

and excellent quality for a long period of time while preventing the problems of early deactivation of the catalyst and colouring of the reaction product.

3.4 From Laboratory-scale to Production: Petrochemicals and Fine Chemicals

The increasing demand for petroleum and the enormous progress in petrochemical industry over the last decades has allowed catalysis by zeolites to reach a high level of maturity in this sector of chemical production.

Continuous flow processes are well implemented in petrochemical industries, where the safe and orderly processing of crude oil into flammable gases and liquids at high temperatures and pressures requires considerable knowledge, control and expertise.

A refinery has a number of plants operating under different conditions producing a suite of categories of petrochemicals (raw chemicals, intermediates and end-products) and the responses to forecast market demands have to be coordinated. Thus, simultaneous planning of total production and each plant production has to be undertaken. Petrochemical industries usually have an efficient model for the different interactions among the plants to provide appropriate production as well as an operation plan for the decision-making process.

In striking contrast, the number of industrial continuous flow production processes within fine chemistry industry is scarce. In general, chemical processing of bulk drugs and their precursors usually takes place in a unique chemical manufacturing environment, where it is not easy to scale up the chemical production for several reasons. The main one is closely related to the great molecular complexity of the products. They are usually polyfunctional molecules with a high molecular weight and lower volatility than bulk chemical products. Besides this, fine chemicals usually have more isomers and lower stability than bulk chemicals.

Another classic problem is related to their synthesis, which is usually carried out in several steps and which takes place in solution under mild reaction conditions (atmospheric pressure and moderate or low temperatures) in small equipment or reactors ($500\,L$ to $10\,m^3$). This means that high value materials are usually transformed in low-volume batch plants that generally operate as multiproduct environments. For example, the actual processes and their operation in an effective manufacturing plant may be organised and carried out reliably on the basis of a relatively modest set of 'unit operations' such as crystallisation, liquid/liquid extraction, filtration and others. These unit operations can be carried out in whatever sequence the unique chemical process requires and, in the multiproduct batch plant, can be usually built as versatile processing modules that can execute one or more of the operations regardless of the specific process at hand.

Another feature of these high added value compounds is their preparation, which is based on classic organic reactions where the catalytic steps are rather scarce, hence leading to high production costs. Nonetheless (in proportion), they generate major economic benefits compared with bulk chemicals. In addition, they have a relatively high E-factor.

Unfortunately only a few new chemicals, obtained through batch operations, can be scaled up directly from the bench to the plant (without demonstration in a pilot plant). This direct scale-up is less time-consuming since it avoids the costs of the pilot plant's design, construction and operation, *etc.* However, not all processes should be scaled up without pilot plant demonstration since in many cases 'surprises' can appear just after the larger scale is developed.

For example, the physical form, purity or performance of the product may change as the process moves from the lab to the plant since there are certain scale-sensitive parameters that cannot be fully tested at the lab scale. Therefore, this staged scale-up strategy is not always practical for specialty chemicals, which are often characterised by multistep batch syntheses and relatively low volume, and where speed to market and rapid ramp-up are essential for commercial success. In addition, the variety of the equipment involved does not facilitate the scale-up process, and the capital invested in space and equipment is high.

A plausible solution to the lack of continuous processes for fine chemical production is the development of continuous batch reactors where the reactants are constantly added at the same time as products are taken out, purified (by distillation, precipitation, filtration, *etc.*) and recycled back to the reactor.

Finally, regarding continuous processing and its potential in pharmaceutical/chemical manufacturing, it has been suggested that processing equipment based on the use of a myriad of microreactors working in parallel could present an effective solution. Microreactors eliminate the problem of costly dedicated plants required by traditional continuous processing. These small continuous microreactors retain the flexibility of batch manufacturing whilst bringing a host of added bonuses such as better selectivity at high temperatures, high throughput and easier energy handling. They are also suitable for fast reactions and reactions with unstable intermediates. This gives the advantage of being able to move quickly from laboratory to pilot plant and manufacture.

Several companies have already taken the initiative and started on the continuous processing path. Lonza, in particular, has thrown its weight behind the technology, working directly with microreactor manufacturers to develop flexible systems that can be used on the multi-kilo and tonne scale, and successfully scaling up some processes.[83]

Pharma companies are also keeping their eyes on those developments as well. Companies such as Organon and AstraZeneca have already begun to use microreactors for some processes and with the recent $65 million hook-up between Novartis and MIT set to 'revolutionise' pharmaceutical production with continuous manufacturing, the sector looks ripe for growth over the coming years.[83]

The French/Swiss company, AETDEV, has also developed reactor technology in this area, with DSM and UK company Phoenix Chemicals also achieving the yield and efficiency improvements that continuous processing technologies can bring.[83]

However, despite the apparent promise continuous processing seems to offer, the new technology still has some way to go before it is sufficiently developed to lure manufacturers away from the zone of batch-based manufacture.

3.5 Future and Industrial Perspectives

In the tendency to replace homogeneously catalysed processes with heterogeneous catalysts, the major contribution of zeolites as catalysts for industrial processes is worthy of note. The main reasons for this are probably their reproducible preparation and shape selectivity. Nonetheless, it is foreseeable that zeolites modified by various ways and methods will make a further contribution to their industrial application.

On the other hand, it is necessary to improve already established industrial processes, because the selectivity and life of catalysts are not necessarily satisfactory in many of them. For this reason, the search for novel solids or modified zeolites that catalyse the transformations with high product selectivity, high reaction rate and low deactivation rate is an ongoing process.

Significant fundamental research is also needed for the development of solid catalysts than can operate in aqueous solution and other green solvents as well as in the absence of solvent.

Thus, multifunctional catalysts are expected to become even more important for industrial applications in future, with the aim to combine several transformations in one step.

With respect to the fine chemicals industry, this is for various reasons (low volumes and high value added products, multistep synthesis, multipurpose batch equipment, short development times, fast commercialisation, *etc.*) less responsive to pressure for changes. It is foreseeable that the experience in the use of zeolites in the petrochemical industry can be increasingly applied to the selective organic synthesis of chemical intermediates and fine chemicals. Synthesis of fine chemicals and specialty chemicals should be raised at the petrochemistry level.

In summary, the search for new materials and associated novel properties will continue in the near future. The demands of society for new technology will necessarily drive the search not only for more new materials but also for faster means of exploiting these discoveries. Therefore, it is expected that advances in computer modelling, analytical instrumentation, experimental technologies (*vs.* high throughput screening and combinatorial methods) and communication networks will all serve to propel new material discoveries and associated new beneficial technologies.

References

1. K. Tanabe and H. Hölderich, *App. Catal. A: Gen.*, 1999, **181**, 399.
2. F. Degnan Jr., *Top. Catal.*, 2000, **13**, 349.
3. N. Y. Chen and T. F. Degnan, *Chem. Eng. Prog.*, 1988, **84**, 33.

4. www.iza-structure.org.

5. (a) I. D. Mikheikin, A. I. Lumpov, G. M. Zhidomirov and V. B. Kazansky, *Kinet. Catal.*, 1978, **19**, 1053; (b) I. D. Mikheikin, I. N. Senchenya, A. I. Lumpov, G. M. Zhidomirov and V. B. Kazansky, *Kinet. Catal.*, 1979, **20**, 496; (c) W. J. Mortier, J. Sauer, J. A. Lercher and H. Noller, *J. Phys. Chem.*, 1984, **88**, 905; (d) W. J. Mortier, in *Proceedings of 6th International Conference on* Zeolites, ed. D. Olson and A. Bisio, Butterworths, Guildford, 1984, **6**, 734; (e) A. Corma, *Chem Rev.*, 1995, **95**, 559.

6. J. B. Uytterhoeven, L. G. Christner and W. K. Hall, *J. Phys. Chem.*, 1965, **69**, 2117.

7. E. M. Flanigen, *Stud. Surf. Sci. Catal.*, 1991, **58**, 1.

8. S. T. Wilson, B. M. Lock, C. A. Mesina, T. R. Cannan and E. M. Flanigen, *J. Am. Chem. Soc.*, 1982, **104**, 1146.

9. B. M. Lock, C. A. Mesina, R. L. Patton, R. T. Gajek, T. R. Cannan and E. M. Flanigen, *J. Am. Chem. Soc.*, 1984, **106**, 6092.

10. E. M. Flanigen, B. M. Lock, R. L. Patton and S. T. Wilson, *Stud. Surf. Sci. Catal.*, 1986, **28**, 103.

11. M. E. Davis, C. Montes, P. E. Hathaway and J. M. Garces, *Stud. Surf. Sci. Catal.*, 1989, **49**, 199.

12. Q. Huo, R. Xu, S. Li, Z. Ma, J. M. Thomas, R. H. Jones and A. M. Chippendale, *J. Chem. Soc. Chem. Commun.*, 1992, **875**.

13. M. Estermann, L. B. McCusker, C. Baerlocher, A. Merrouche and H. Kessler, *Nature*, 1991, **352**, 320.

14. (a) H. Hattori, *Chem Rev.*, 1995, **95**, 537 (and references herein); (b) P. E. Hathaway and M. E. Davis, *J. Catal.*, 1989, **116**, 263; (c) C. B. Dart and M. E. Davis, *Catal. Today*, 1994, **19**, 151; (d) F. Yagi, N. Kanuka, H. Tsuji, H. Kita and H. Hattori, *Stud. Surf. Sci. Catal.*, 1994, **90**, 349.

15. (a) J. S. Beck, C. T. W. Chi, I. D. Johnson, C. T. Kresge, M. E. Leonowicz, W. J. Roth and J. C. Vartuli, WO Pat. 91/11390 (1991); (b) C. T. Kresge, M. E. Leonowicz, W. J. Roth, J. C. Vartuli and J. S. Beck, *Nature*, 1992, **359**, 710; (c) J. S. Beck, J. C. Vartuli, W. J. Roth, M. E. Leonowicz, C. T. Kresge, K. D. Schmitt, C. T. W. Chu, D. H. Olson, E. W. Sheppard, S. B. McCullen, J. B. Higgins and J. L. Schlenker, *J. Am. Chem. Soc.*, 1992, **114**, 10834.

16. A. Corma, M. J. Diaz-Cabanas, J. L. Jorda, C. Martinez and M. Moliner, *Nature*, 2006, **443**, 842.

17. (a) W. Schoonover and M. J. Cohn, *Top. Catal.*, 2000, **13**, 367; (b) W. M. Meier, D. H. Olson and Ch. Baerlocher, *Atlas of Zeolite Structure Types,* Elsevier, London, 4th rev edn. 1996.

18. A. Corma, *Stud. Surf. Sci. Catal.*, 1994, **83**, 461.

19. Ch. Baerlocher, L. B. McCusker and D. H. Olson, *Atlas of Zeolite Framework Types,* Elsevier, Amsterdam, 6th rev edn, 2007.

20. (a) I. E. Maxwell, *Stud. Surf. Sci. Catal.*, 1991, **58**, 12; (b) C. Marcilly, *Stud. Surf. Sci. Catal.*, 2001, **135**, 37; (c) W. Vezmeizen and J. P. Gilson, *Top. Catal.*, 2009, **52**, 1131.

21. F. Degnan Jr., *Top. Catal.*, 2000, **13**, 349.
22. N. Y. Chen and T. F. Degnan, *Chem. Eng. Prog.*, 1988, **84**, 33.
23. N. Y. Chen, W. E. Garwood and F. G. Dwyer, *Shape Selective Catalysis in Industrial Applications,* Marcel Dekker, New York, 1996.
24. I. E. Maxwell, *CATTECH*, 1997, **1**, 5.
25. (a) S. T. Sie, *Stud. Surf. Sci. Catal.*, 1994, **58**, 17; (b) P. R. Pujado, J. A. Rabo, G. J. Antos and S. A. Gembicki, *Catal. Today*, 1992, **13**, 113.
26. A. Corma, *Catal. Lett.*, 1993, **22**, 33.
27. N. A. Cusher, in *Handbook of Petroleum Refining Processes*, ed. R. A. Myers, McGraw-Hill, New York, 2003, p. 9.15.
28. J. Weitkamp, in *Hydrocracking and Hydrotreating*, ed. J. W. Ward and G.A. Quader, American Chemical Society, Washington DC, ACS Symposium Series 20, 1975, p. 1.
29. J. Lázaro Múñoz, A. Corma, and J. M. Frontela Delgado, *Span. Pat.*, ES 2014811, 1990; *US Pat.*, US5 057 471 A3, 1990.
30. A. Corma, J. Frontela, J. Lazaro and M. Perez, *Prepr. – Amer. Chem. Soc., Div. Pet. Chem.*, 1991, **36**, 833.
31. M. Hino and K. Arata, *Chem. Lett.*, 1979, **10**, 1259.
32. K. Arata, *Appl. Catal. A*, 1996, **146**, 3.
33. N. Y. Chen, J. Maziuk, A. B. Schwartz and P. B. Weisz, *Oil Gas J.*, 1968, **66**, 154.
34. N. Y. Chen, US Pat., US3.729.409, 1973.
35. W. Zhang and P. G. Smirniotis, *J. Catal.*, 1999, **182**, 400.
36. A. de Lucas, J. L. Valverde, P. Sanchez, F. Dorado and M. J. Ramos, *Appl. Catal. A*, 2005, **282**, 15.
37. A. de Lucas, P. Sánchez, F. Dorado, M. J. Ramos and J. L. Valverde, *Appl. Catal. A*, 2005, **294**, 215.
38. N. P. Kuzsnetsov, *J. Catal.*, 2003, **218**, 12.
39. G. Kinger and H. Vinek, *Appl. Catal., A*, 2001, **218**, 139.
40. J. M. Campelo, F. Lafont and J. M. Marinas, *J. Catal.*, 1995, **156**, 11.
41. J. M. Campelo, F. Lafont and J. M. Marinas, *Appl. Catal., A*, 1998, **170**, 139.
42. S. J. Miller, *Micropor. Mater.*, 1994, **2**, 439.
43. A. Corma, P. J. Miguel and A. V. Orchilles, *Appl. Catal. A*, 1994, **117**, 29.
44. T. R. Hughes, W. C. Buss, R. W. Tamm and R. L. Jacobson, *Stud. Surf. Sci. Catal.*, 1986, **28**, 725.
45. N. Y. Chen, R. L. Gorring, H. R. Ireland and T. R. Stein, *Oil Gas J.*, 1977, **75**, 165.
46. J. Walendziewski and B. Pniak, *Appl. Catal. A*, 2003, **250**, 39.
47. B. W. Wojcciechowski and A. Corma, *Catalytic Cracking,* Marcel Dekker, New York, 1986.
48. J. Biswas and I. E. Maxwell, *Stud. Surf. Sci. Catal.*, 1989, **49**, 1263.
49. J. M. Maselli and A. W. Peter, *Catal. Rev.*, 1984, **26**, 525.
50. I. E. Maxwell, *Catal. Today*, 1985, **1**, 385.
51. J. A. R van Veen, J. K. Minderhoud, L. G. Huve and W. H J. Stork, in *Handbook of Heterogeneous Catalysis*, ed. E. Gerhard, H. Knözinger, F. Schüth and J. Weitkamp, Wiley, New York, 2008, **Vol. 6**, p. 2778.

52. A. Sequeira Jr., *Lubricant Base Oil and Wax Processing*, Marcel Dekker, New York, 1994.

53. F. A. Smith and R. W. Bortz, *Oil Gas J.*, 1990, **13**, 50.

54. S. J. Miller, *US Pat.*, US4,859,311, 1989.

55. T. Blasco, A. Chica, A. Corma, W. J. Murphy, J. Agundez-Rodriguez and J. Perez-Pariente, *J. Catal.*, 2006, **242**, 153.

56. (a) C. R. Marcilly, *Top. Catal.*, 2000, **13**, 357; (b) J. Cejka and B. Wichterlova, *Catal. Rev.*, 2002, **44**, 375.

57. J. A. Martens, J. Perez-Pariente, E. Sastre, A. Corma and P. A. Jacobs, *Appl. Catal.*, 1988, **45**, 85.

58. M. Guisnet, N. S. Gnep and S. Morin, *Micropor. Mesopor. Mater.*, 2000, **35–36**, 47.

59. F. Bauer, W. H. Chen, H. Ernst, S. J. Huang, A. Freyer and S. B. Liu, *Micropor. Mesopor. Mater.*, 2004, **72**, 81.

60. F. Moreau, S. Bernard, N. S. Gnep, S. Lacombe, E. Merlen and M. Guisnet, *J. Catal.*, 2001, **202**, 402.

61. (a) R. A. Sheldon, *Chemtech.*, 1994, March, 38; (b) A. Mitsutani, *Catal. Today*, 2002, **73**, 57.

62. (a) T. Okuhara, *Chem. Rev.*, 2002, **102**, 3641; (b) P. M. M. Blauwhoff, J. W. Gosselink, E. P. Kieffer, S. T. Sie and W. H. J. Stork, in *Catalysis and Zeolites*, ed. J. Weitkamp and L. Puppe, Springer, Berlin, 1999, p. 437; (c) W. F. Hölderich, J. Röseler, G. Heitmann and A. T. Liebens, *Catal. Today*, 1997, **37**, 353.

63. (a) H. Ishida, Y. Fukuoka, O. Mitsui and M. Kono, *Stud. Surf. Sci. Catal.*, 1994, **83**, 473; (b) P. B. Venuto, *Micropor. Mater.*, 1994, **2**, 297.

64. (a) H. Sato, K. Hirose, N. Ishii and Y. Umada, *US Pat.*, 4 709 024, 1986, Sumitomo Chemical Co. Ltd; (b) H. Sato, K. Hirose, M. Kitamura, Y. Umada, N. Ishii and H. Tojima, *US Pat.*, 4 717 769, 1987, Sumitomo Chemical Co. Ltd; (c) L. Marosi, M. Schwarzmann and J. Stabenow, *Eur. Pat.*, EP7081, 1979, BASF AG; (d) H. Ichihashi and M. Kitamura, *Catal. Today*, 2002, **73**, 23; (e) H. Ichihashi and H. Sato, *Appl. Catal,. A: Gen.*, 2001, **221**, 359.

65. M. Kitamura and M. Shimazu, *JP* 2000-229939, 1999, Sumitomo Chemical Co. Ltd.

66. K. Eguchi, T. Tokiai and H. Arai, *Appl. Catal.*, 1987, **34**, 275.

67. H. Ishida and K. Akagishi, *Nippon Kagaku Kaishi*, 1996, **290**.

68. S. Namba, S. Nakanishi and T. Yashima, *J. Catal.*, 1984, **88**, 505.

69. (a) J. Kaminska, M. A. Schwegler, A. Hoefnagel and H. van Bekkum, *Rec. Trav. Chim. Pays-Bas*, 1992, **111**, 432; (b) M. G. Clerici, *Top. Catal.*, 2000, **13**, 373.

70. Firmenich S.A., *Eur. Pat.*, EP155591, 1988.

71. A. Severino, A. Esculcas, J. Rocha, J. Vital and L. S. Lobo, *Appl. Catal., A*, 1996, **142**, 255.

72. P. J. Kunkeler, J. van der Waal, J. Bremmer, B. J. Zuurdeed, R. S. Dowing and H. van Bekkum, *Catal. Lett.*, 1998, **53**, 135.

73. (a) M. Spagnol, L. Gilbert, R. Jacquot, H. Guillot, P. J. Tirel and A. M. I. Govic, Poster presented at 4th International Symposium on Heterogeneous Catalysis and Fine Chemicals, 1996; (b) M. Spagnol, L. Gilbert, H. Guillot and P. J. Tirel, *WO* 97/48665, 1997; (c) S. Ratton, *Chem. Today*, 1997, **3–4**, 33; (d) M. Spagnol, L. Gilbert and D. Alby, in *The Roots of Organic Development*, ed. J. R. Desmurs and S. Ratton, Elsevier, Amsterdam, 1996 Industrial Chemistry Library, Vol. 8, p. 29.
74. D. Rohan, C. Canaff, E. Fromentin and M. Guisnet, *J. Catal.*, 1998, **177**, 296.
75. G. Harvey, A. Vogt, H. W. Kouwenhoven and R. Prins, in *Proceedings of the 9th International Zeolite Conference*, ed. R. von Ballmoos, J. B. Higgins and M. M. J. Treacy, **Vol. II**, Butterworth-Heinemann, Boston, 1993, p. 363.
76. M. Spagnol, L. Gilbert, E. Benazzi and C. Marcilly, *US Pat.*, 5 817 878, 1998, Rhône Poulenc.
77. A. J. Hoefnagel and H. van Bekkum, *Appl. Catal.*, A, 1993, **97**, 87.
78. H. van Bekkum, A. J. Hoefnagel, M. A. Vankoten, E. A. Gunnenwegh, A. H. G. Vogt and H. W. Kouwenohoven, *Stud. Surf. Sci. Catal.*, 1994, **83**, 379.
79. (a) C. B. Dartt and M. E. Davis, *Catal. Today*, 1994, **19**, 151; (b) C. B. Dartt and M. E. Davis, *Ind. Eng. Chem. Res.*, 1994, **33**, 2887.
80. (a) F. P. Gortsema, J. J. Sharkey, G. T. Wildman and B.S. Beshty, *Eur. Pat.*, 0481674, 1992; (b) F. P. Gortsema, B. Beshty, J. J. Friedman, D. Matsumoto, J. J. Sharkey, F. Wildman, T. J. Blacklock and S. A. Pan, in *Proceedings of 14th Conference on the Catalysis of Organic Reactions*, Albuquerque, NM, 1992.
81. K. Nagai, *Applied Catal. A. Gen.*, 2001, **221**, 367.
82. Y. Shima, T. Abe and H. Higuchi, *Eur. Pat.*, 0598243, 1994, Mitshubishi Gas Chemical Co. Ltd.
83. http://www.in-pharmatechnologist.com/Processing-QC/Ditch-batch-continuous-is-the-future, accessed on July 2009.

Microfluidic Devices for Organic Processes

PAOLA LAURINO, ARJAN ODEDRA,[‡] XIAO YIN MAK,
TOMAS GUSTAFSSON,* KAROLIN GEYER[†] AND
PETER H. SEEBERGER

Max Planck Institute of Colloids and Interfaces, Department of Biomolecular
Systems, Research Campus Golm, D-14424, Potsdam, Germany

4.1 Microreactors and Microfluidic Devices: Concepts and Definitions

Synthetic organic chemists have long relied on the use of the 'round-bottomed flask' as a means for carrying out organic reactions. Indeed, the reaction flasks found in laboratories worldwide today are not very different from those used in the 19th century. Recently, microfluidic devices have emerged as a new technology to facilitate organic transformations and hold some advantages over the use of traditional reaction vessels. Continuous flow devices have been used in other fields for some time and the potential for adopting this technology in organic synthesis is being realised. Interest in miniaturised chemical and biological systems has grown tremendously. The development of this technology has created new chemical engineering challenges as well as many novel research applications for synthetic organic chemists.[1]

[‡]Present address: Pharmacenter, University of Basel, Kingelbergstrasse 50, CH 4056, Basel, Switzerland
*Present address: Medicinal Chemistry, AstraZeneca R&D Mölndal, SE 43183 Mölndal, Sweden
[†]Present address: Davenport Chemical Laboratories, Department of Chemistry, University of Toronto, 80 St. Georg Street, Toronto, M5S 3H6 Ontario, Canada

---

RSC Green Chemistry No. 5
Chemical Reactions and Processes under Flow Conditions
Edited by S.V. Luis and E. Garcia-Verdugo
© The Royal Society of Chemistry 2010
Published by the Royal Society of Chemistry, www.rsc.org

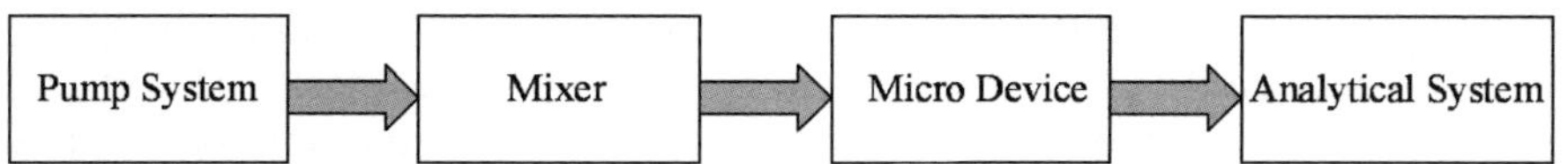

Figure 4.1 Schematic representation of an integrated continuous microfluidic device.

Continuous microflow reactors are defined as miniaturised systems consisting of a channel network with dimensions usually ranging from a few to several hundred micrometres in scale. This network of channels is often embedded in the flat surfaces of materials such as silicon, glass, stainless steel, ceramics or polymers. The most common material used for chemical application is glass, which is relatively inert. Reagents are flushed into these systems in a pre-determined sequence *via* pumps – syringe pumps, high-performance liquid chromatography (HPLC) pumps or peristaltic pumps – or non-mechanical techniques (electrokinetic pumping).[2] Reagents pass through a mixing zone, which can significantly influence the successful implementation of the micro-reactor system as a whole. Finally, continuous flow reactors can be integrated with an on-line analytical system such as HPLC, allowing for immediate control and optimisation of chemical reactions (Figure 4.1).

The applicability of a microfluidic device for a given chemical transformation depends strongly on the material, channel geometry, and the type of pumping mechanism used to drive and regulate the flow. The material has to be chemically inert, temperature and electrically stable, and ideally transparent. Reaction temperature, pressure and solvent viscosity should be taken into consideration, in conjunction with the type of pumping mechanism, to ensure a reproducible and homogeneous flow. Design of the channel geometry has to ensure efficient mixing as well as a reproducible retention times.[3]

Numerous reactor types have been developed and applied to synthetic transformations. Some of these systems were used for synthetic transformations on a lab scale, others for production purposes. However, all mentioned devices generally have two main common properties:

- **Small channel diameters** provide an increased surface-to-volume ratio and allow for excellent control of heat and mass transfer of chemical reactions. This parameter can influence the reaction time, yield and selectivity of chemical transformations.
- **Laminar flow** behaviour and short diffusion paths in the channels ensure efficient and rapid mixing.

4.2 Main Advantages of Microfluidic Devices

Numerous advantages associated with the use of microreactors have rendered this technology a potentially important tool for organic chemists. Faster, easier and reproducible chemistry can be performed in microfluidic devices when compared with the traditional round-bottomed flask.[4]

The heat transfer coefficient is usually several orders of magnitude higher for a microstructured device compared with a round-bottomed flask.[5] This effect is largely due to the greatly increased surface-to-volume ratio of microchannelled reactors. The increased heat transfer rate not only allows for precise temperature control but also serves to reduce the occurrence of a non-uniform temperature profile inside the reactor. In addition, reactions can often be performed at much higher temperatures than in batch, where cooling is more efficient on the surface than in the centre of the reaction vessel.

Reactions carried out in traditional equipment cannot always be performed at optimal temperatures and hence are often slower. Microfluidic devices, on the other hand, can sustain these temperatures due to its high heat transfer capability. The reactants are fed continuously into, and removed from, the reactor, thereby only exposing them to the required reaction conditions for the shortest possible time. This procedure constitutes a significant advantage when forcing conditions (high temperature, high pressure, corrosive reagents, *etc.*) are used or when sensitive intermediates are formed. Removal of products from the reaction zone, to be further reacted or else quenched to form less sensitive substrates, occurs as soon as conversion of reactants is complete.[3]

Fast equilibration can occur within the cross-section of the reactor through fast diffusion controlled mass transfer as a consequence of the small channel dimensions of the reactor. Uniform reaction conditions are maintained over time leading to overall increased reproducibility. The internal reaction volume is constant, and temperature and flow rates can be precisely controlled throughout the reactor.

Microreactors generally lend themselves well to pressurisation compared with conventional reaction flasks. This enables the superheating of solvents (*i.e.* heating above the boiling point), thus removing the boiling point barrier for the amount of energy that can be supplied to the reaction. Increased temperatures generally lead to faster kinetics and, consequently, reaction time can be shortened considerably. Gases dissolve more easily into the solution through pressurisation, allowing gaseous reagents to be used readily.[6]

Another attractive feature of employing microreactor technology is that only a small inventory of reactants is retained inside the reactor at any given time. From the perspective of operational safety, intermediates or products that are sensitive, high in energy or prone to auto-catalytic decomposition can be safely reacted inside microreactors at much lower risk. The small reactor size simplifies and speeds up the optimisation process, as conditions are easy to adjust and only small amounts of starting materials are required.

4.3   Scale-up of Microflow Reactions

Steady-state reaction conditions, which are reached inside the microflow reactors, are the same regardless of the amount of reactants—an aspect that is rarely achieved with batch reactors. As a consequence of the continuous mode of operation, reactions performed in microreactors can be adjusted from small- to large-scale by simply prolonging the introduction of reagents and allowing for a longer total processing time (Figure 4.2a) or by running several reactors in parallel—'numbering-up' (Figure 4.2b).[7] However the latter method is often less practical since it requires several reactors to be maintained at the same temperature and pressure. Difficulties also arise from the need to deliver reactants uniformly to all reactors, requiring multiple pumps or precise flow dividers.[8]

Microwave heating has significantly impacted organic chemistry.[9] However, the scalability of microwave technology is limited.[10] The continuous flow systems are attractive alternatives, since high temperatures and pressures similar to those produced in microwave reactors can be sustained. In addition, the

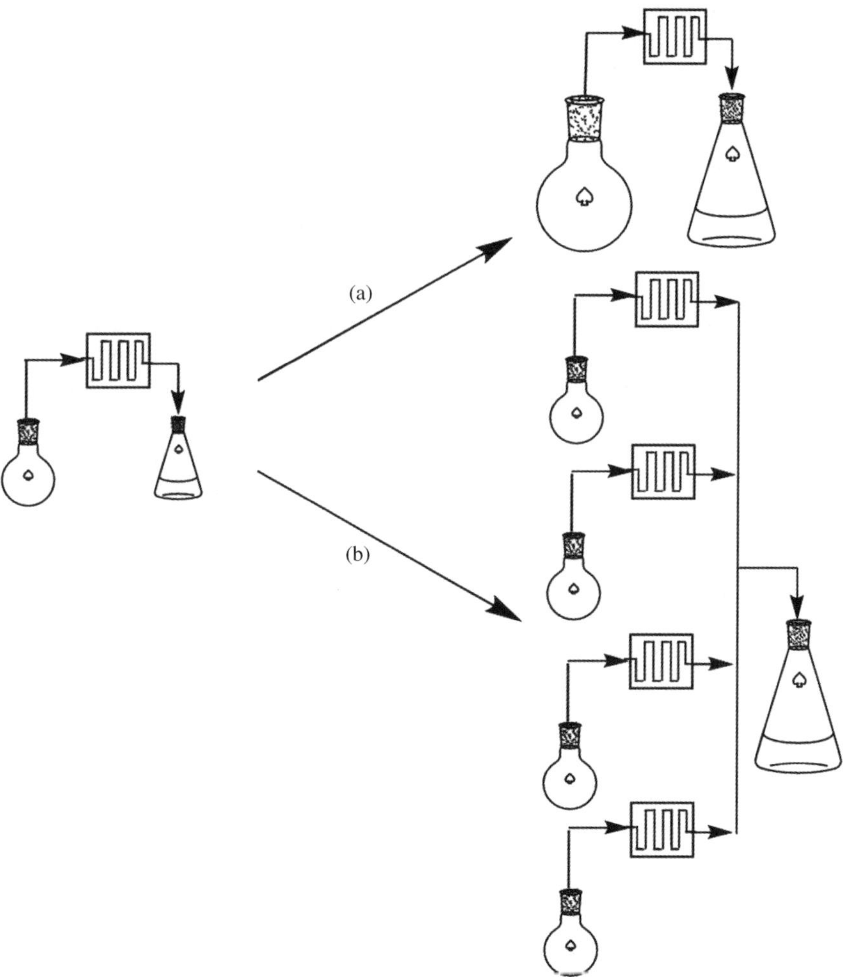

Figure 4.2 Continuous flow approaches to large-scale production.

combination of microwave technology with continuous flow has provided further opportunities for scalable microwave assisted synthesis.[11]

The subsequent sections in this chapter focus on selected examples of microreactor technology in organic synthesis, highlighting the broad range of organic transformations that have been explored using continuous flow, as well as the advantages gained in the use of these microdevices.

4.4 Liquid–Liquid Reactions

4.4.1 Photochemical Reactions

Photochemical reactions are fundamental to organic synthesis,[12] but are difficult to scale-up. The efficiency of irradiation is difficult to maintain, *e.g.* maintaining a consistent surface area of irradiation to reaction volume. Long irradiation times are often required and can lead to increases in the formation of by-products or the decomposition of starting materials. The use of continuous microflow devices for photochemistry allows for consistent and uniform irradiation of the reaction. Light penetration is maximised due to the small channel depths of the reactors. Residence times can be reduced, and the problems typically encountered using standard laboratory reaction vessels can be avoided.[13]

Fukuyama *et al.* investigated the [2 + 2] cycloadditions of vinyl acetates with various cyclohexenone derivatives using a microflow system consisting of a glass microreactor (1000 μm width and 500 μm depth) and a 300 W high pressure Hg lamp.[14] Yields were improved and reaction times were significantly reduced compared with the same reaction run in batch using a Pyrex flask. Two serially connected micro-photoreactors used in tandem at twice the flow rate used for a single reactor provided a similar result (Scheme 4.1).

Device	Residence time (h)	Yield (%)
microreactor	2	88
two serially connected microreactors	2	85
pyrex flask	2	8
pyrex flask	2	22

Scheme 4.1 [2 + 2] Cycloaddition of cyclohexenone and vinyl acetate.

In addition to their studies of [2 + 2] cycloadditions, Fukuyama *et al.* investigated the use of a Barton reaction to construct steroid **5**, an important intermediate in the synthesis of an endothelial receptor antagonist (Scheme 4.2).[15] Initial optimisation studies of this reaction involved the comparison of various glass covers (quartz, Pyrex and soda lime glass) for the stainless steel microreactor, as well as examination of the light source (300 W high-pressure Hg lamp, or 15 W black light lamp) and its distance from the microreactor. A gram-scale synthesis was accomplished under optimal conditions using two serially connected microreactors with a combined of volume of 8 mL, a residence time of 32 min and irradiation by eight 20 W black light lamps. After 20 hours of operation, 3.1 g of the desired product **5** were isolated (60% yield).

Another example of a photo-cycloaddition was reported by Booker-Milburn *et al.* using a FEP (fluorinated ethylene propylene) flow reactor[16] and a Pyrex-filtered 400 W medium-pressure Hg Lamp.[17] The intramolecular [5 + 2] cycloaddition of **6** was carried out under flow conditions, affording **8**, a key intermediate in the synthesis of the *Stemona* alkaloid (±)-neostenine, in 63% yield (Scheme 4.3). By employing a high flow rate and a powerful lamp, photo-degradation of the product was minimised and up to 1.3 g of the cycloadduct were prepared in a single nine hour run, demonstrating the effectiveness of the system for scale-up. By comparison, previous attempts in a batch photo-chemical reactor provided **8** at scale greater than 100 mg in less than 20% yield.

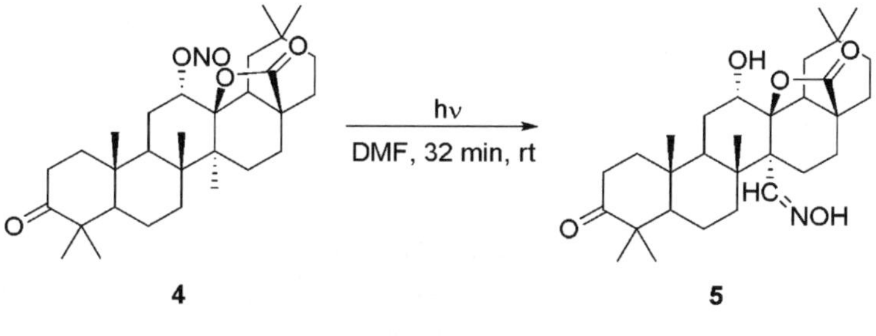

Scheme 4.2 Barton nitrite photolysis of steroid **5**.

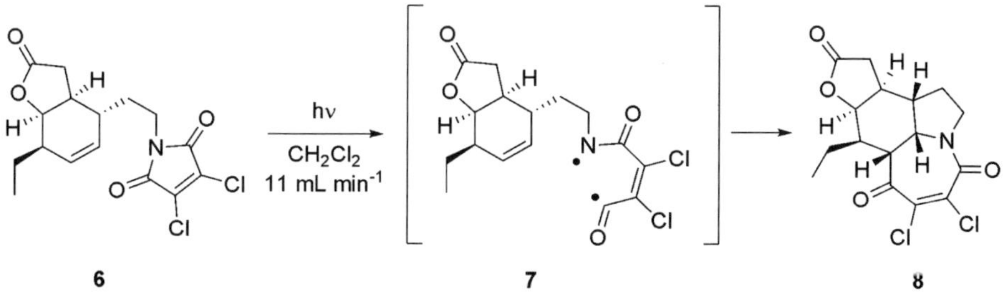

Scheme 4.3 Key step in the synthesis in (±)-neostenine.

4.4.2   Heterocycle Synthesis

Continuous microflow devices have been successfully applied to the synthesis of heterocycles. Acke and Stevens reported on the multicomponent synthesis of isochromen-1-one derivatives using a stacked plate microreactor combined with a residence time unit (tubing) (Scheme 4.4).[18] Direct handling of HCN, a hazardous and toxic reagent required for the synthesis, was avoided by *in situ* generation in the microreactor using potassium cyanide and acetic acid.

Use of the microreactor permitted a lower reaction temperature compared with the elevated temperatures required in batch. Crystallisation of the iso-chromen-1-one products on the reactor wall resulted in blockage of the tubing towards the end of the reactor system, requiring higher dilution. Alternatively, reaction mixture plugs were created and transported by a pumped flow of an immiscible solvent, Fluorinert FC-70, permitting the use of higher concentrations. Under these conditions, crystallisation was prevented by the lack of contact of the reaction plugs with the reactor walls.

The synthesis of 1,2-pyrazoles *via* electroosmotic flow was demonstrated by Watts *et al.* using a borosilicate glass microreactor (Scheme 4.5).[19] The yields were considerably higher when compared with the corresponding batch synthesis. This improvement was likely to be the effect of improved temperature control within the microflow device.

This flow technique was further used for the preparation of isoxazole **14** and substituted pyrazole **15**, as shown in Scheme 4.6. In both reactions, the yields of the products were significantly higher than in batch. In the case of isoxazole **14**, the yield was improved from 52% to 98%; similarly, the yield of benzylated pyrazole **15** was increased from 76% to 100%.

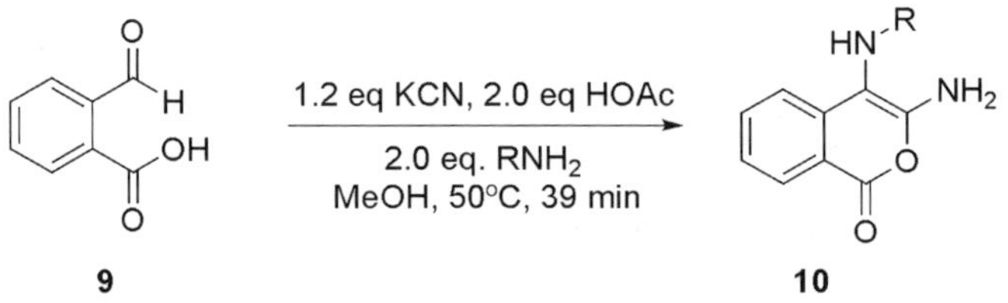

RNH$_2$	Yield (%)	Output (g/h)
Aniline	66	1.80
3-Methoxyaniline	75	2.28
3-Methylaniline	69	1.98
N-Methylaniline	49	1.41
Allylamine	6	0.14

Scheme 4.4   Synthesis of isochromen-1-one derivatives.

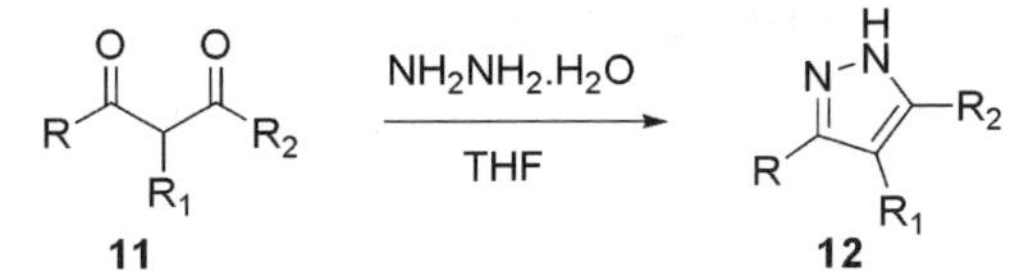

R	R$_1$	R$_2$	Conversion (%)^a in Batch	Conversion (%)^a in Microreactor
CH$_3$	H	CH$_3$	62	100
Ph	H	CH$_3$	63	100
-(CH$_2$)$_4$-		Ph	72	100
Ph	H	Ph	64	100
Ph	CH$_3$	CH$_3$	71	98

^{a}Conversion determined by GC-MS analysis

Scheme 4.5 Synthesis of 1,2-pyrazole derivatives.

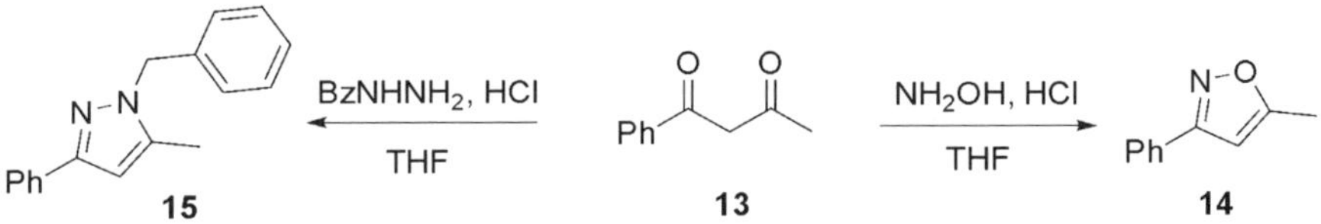

Scheme 4.6 Synthesis of isoxazole **14** and pyrazole **15**.

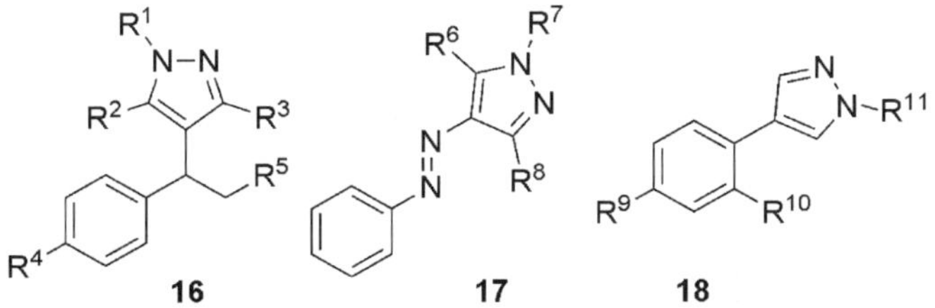

Figure 4.3 Pyrazole library.

Garcia-Egidio *et al.* also developed a microreactor system for the synthesis of pyrazoles, coupled with simultaneous on-line analysis.[20] A combinatorial library of 21 different pyrazoles was synthesised (Figure 4.3) without cross-contamination using a plug flow approach, whereby different reagent and solvent pulses were pumped continuously through the reactor. Only small amounts of starting material were required due to the small device, with detection and analysis of the products performed directly *via* an on-line LC-UV-MS (liquid chromatography–ultraviolet–mass spectrometry) system.

4.4.3 Synthesis of Bio-oligomers

The synthesis of bio-oligomers such β-peptides and oligosaccharides was described by the Seeberger laboratory.[21] β-Tetrapeptides were prepared in a silicon microreactor, enabling the rapid survey of reaction parameters to identify optimal reaction conditions (Scheme 4.7). The peptide tetramer **23** was synthesised using this microreactor strategy with yields better than those obtained using standard solution and solid phase procedures, albeit with significantly shorter reaction times. The shortened reaction times, in combination with higher than usual reaction temperatures, prevented product precipitation during the reaction. In batch mode, this precipitation usually generated a gel-like reaction mixture, leading to poor mixing and longer reaction times. Use of a fluorous benzylic ester protecting group on the first amino acid **19** greatly simplified the downstream purification of the β-peptides.

Microreactor systems also have been applied to the synthesis of oligosaccharides.[22] The synthesis of a glucopyranoside tetramer using the Fmoc-protected glucosyl phosphate building block **28** was optimised and scaled up at ambient temperature, in a silicon-glass microreactor (Scheme 4.8).[22d] Purification of the glycosylation products by fluorous solid phase extraction was

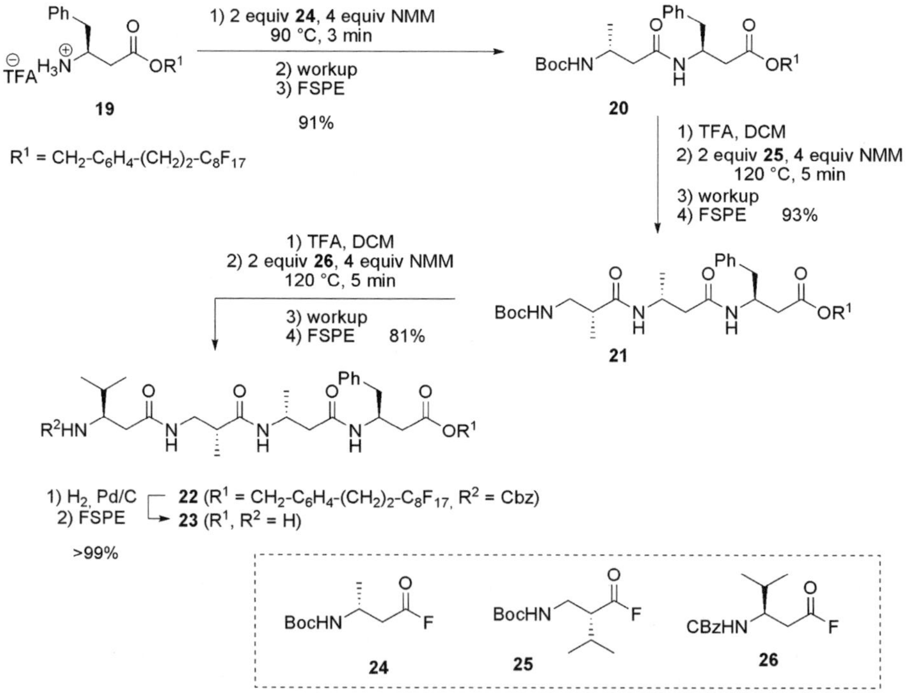

Scheme 4.7 Synthesis of tetrapeptide derivatives **22** and **23** (FSPE = fluorous solid phase extraction). Reactions were monitored via LC-MS analysis, with Fmoc-β³hPhe-OBn as internal standard (Fmoc = fluorenylmethyloxycarbonyl; NMM = μ-methylmorpholine).

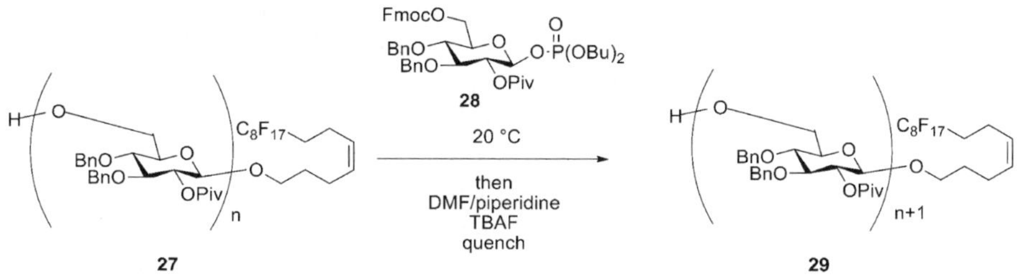

Nucleofile	Solvent	Donor 28 (equiv)	TMSOTf (equiv)	Time (sec)	TBAF (equiv)	Yield(%)
n = 0	TFT	2	2	30	1.5	99
n = 1	CH$_2$Cl$_2$	2	2	20	1.5	97
n = 2	CH$_2$Cl$_2$	3	3	60	2	90
n = 3	CH$_2$Cl$_2$	3	3	60	2	95

Scheme 4.8 Microreactor synthesis of oligosaccharides. Reactions were monitored by LC-MS analysis with 2,3,4,6–tetra-O-benzyl-α-D-glucopyranoside as internal standard.

facilitated by the incorporation of a fluorinated linker system. The initial monoglycoside acceptor ($n = 0$) was not soluble in dichloromethane (DCM), thus requiring the use of trifluorotoluene (TFT).

4.4.4 Multistep Reactions

Rapid access to target compounds can be gained by adapting multistep synthetic sequences to flow conditions. Fast optimisation, as well as significant

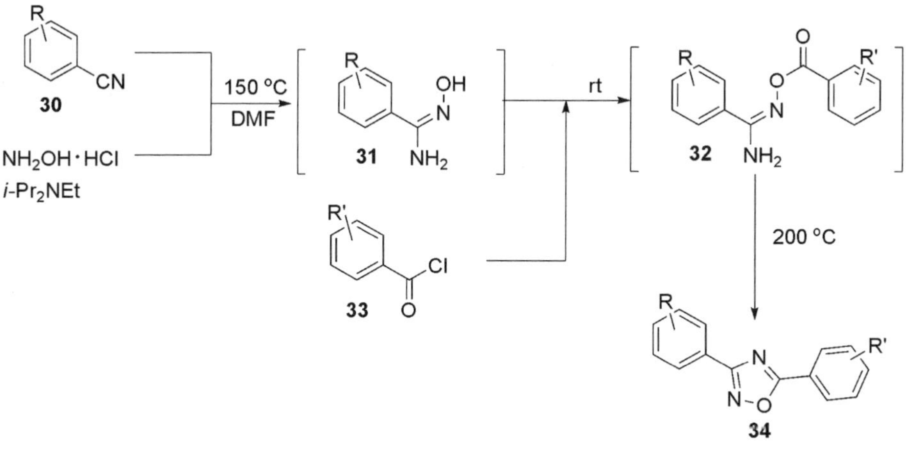

Scheme 4.9 Continuous synthesis of 1,2,4-oxadiazoles.

reductions in reaction time, can be achieved and laborious post-reaction processes such as the purification and isolation of intermediates can often be avoided by conducting several reaction steps consecutively in flow. For example, a multistep synthesis of bis-substituted 1,2,4-oxadiazoles **34** was reported recently by Cosford *et al.*[23] which relied on the use of three sequential micro devices (Scheme 4.9). The superheating of solvent was feasible in the microreactors used, significantly shortening reaction times. The overall time required for the synthesis of the 1,2,4-oxadiazoles was drastically reduced from three days to about 35 min.

An impressive synthesis of a molecular imaging probe, 2-deoxy-2-[^{18}F]fluoro-D-glucose was achieved using microreactor technology. Five different stages of the process including [^{18}F] fluoride concentration, water evaporation, radio-fluorination, solvent exchange and hydrolytic deprotection were integrated into a single microfluidic device.[29] The [^{18}F] radiolabel has a short half-life (110 min) and so rapid synthesis of the probe is essential. Using the microdevice, it was obtained with high radiochemical yield and purity, and in a shorter reaction time compared to conventional methods (14 min *vs.* 50 min).

Continuous flow devices in series can also be used to separate the individual stages of a reaction. This can be useful, as optimal conditions can be developed for each step of a given process. A two-step process for the synthesis of 6-hydroxybuspirone **37** was developed and scaled up using a continuous flow system.[30] Continuous enolisation of buspirone **35** proceeded in a cooled reactor system to form enolate **36**, which was then prepared into a second reactor and oxidised using a counter-current flow of oxygen gas to give **37** (Scheme 4.10). Using continuous flow, only small amounts of the potentially explosive peroxy intermediate was formed at any one time in the contained system. Reaction times were significantly shortened compared to the batch process due to the improved mass transfer of oxygen.

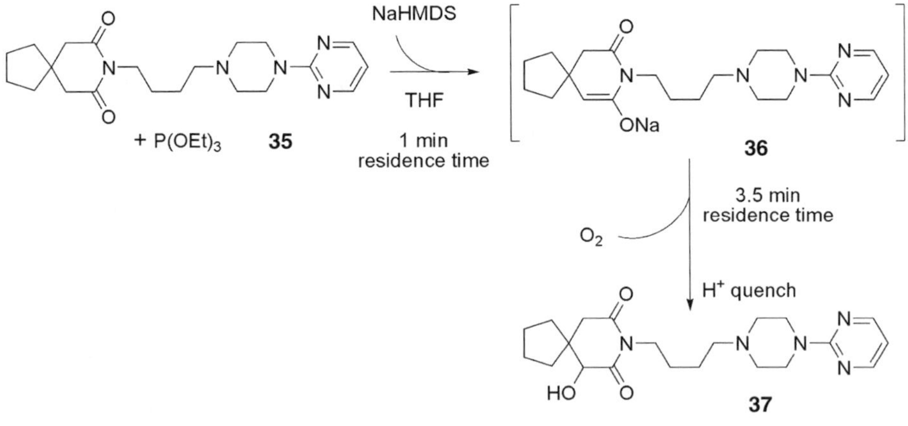

Scheme 4.10 Preparation of 6-hydroxybuspirone **37**.

4.4.5 Free Radical Reactions

Radical reactions are extremely versatile with good functional group tolerance, and are becoming increasingly popular in organic synthesis.[24] However, radical reactions are often difficult to conduct in a controlled manner. Dithiocarbonate derivatives and halogen compounds have been reduced to the corresponding alkanes in a microreactor using tris(trimethylsilyl)silane (TTMSS) as the reducing agent.[25] Reactions were completed within 5 min at 130 °C in toluene [Scheme 4.11, eqn (1)]. The hydrosilylation of alkynes and olefins were carried out under similar conditions to afford the corresponding silylated products in very good yields. Improved Z/E ratios were obtained for the hydrosilylation of alkynes when compared to batch reactions [Scheme 4.11, eqn (2)].

Ryu and co-workers reported the dehalogenation of various organic halides with tributyltin hydride as reductant (Scheme 4.12).[26] By a judicious choice of

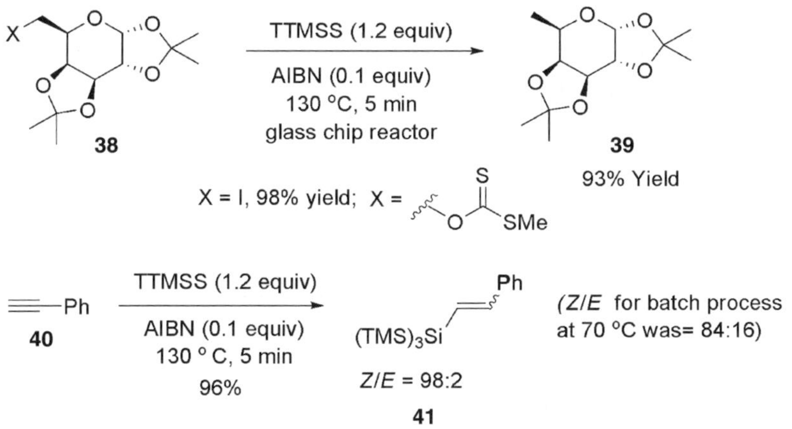

Scheme 4.11 Free radical reactions in the microreactor [AIBN = 2,2′-azobis(2-methyl-propionitrile)].

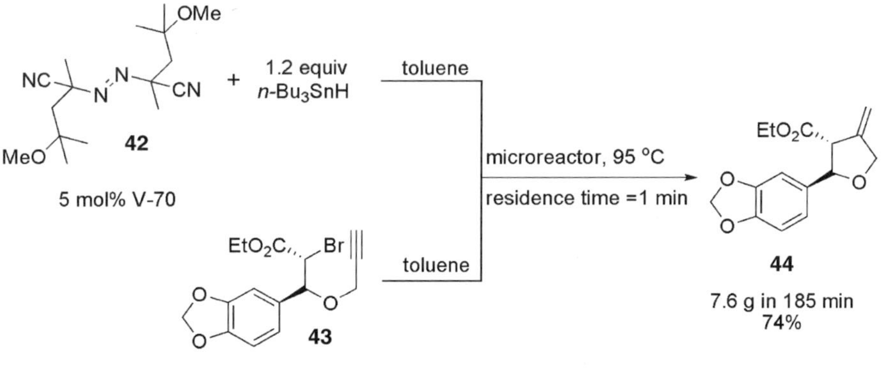

Scheme 4.12 Free radical cyclisation reaction in the microreactor.

the radical initiator, it was possible to reduce various aliphatic bromides within one minute at 80 °C. This methodology was extended to the gram-scale radical cyclisation reaction of substrate **43** to provide tetrahydrofuran derivative **44**, an important intermediate for the synthesis of paulownin[27] and samin.[28]

4.4.6 Reactions Involving Hazardous Materials and Unstable Intermediates

4.4.6.1 *Metal–Halogen Exchange Reactions*

Metal–halogen exchange reactions are usually fast and highly exothermic. Slow addition at low temperature is essential in order to avoid rapid temperature increases when the reaction is carried out in a conventional flask reactor. The small dimensions and high surface to volume ratio of microreactors are thus well-suited towards extremely fast and exothermic reactions, and for the control of highly reactive, short-lived reactive intermediates.

The halogen–magnesium exchange between C_2F_5I and MeMgCl, followed by reaction with benzophenone, has been investigated [Scheme 4.13, eqn (1)].[31] The halogen–magnesium exchange was found to be completed within 0.9 min

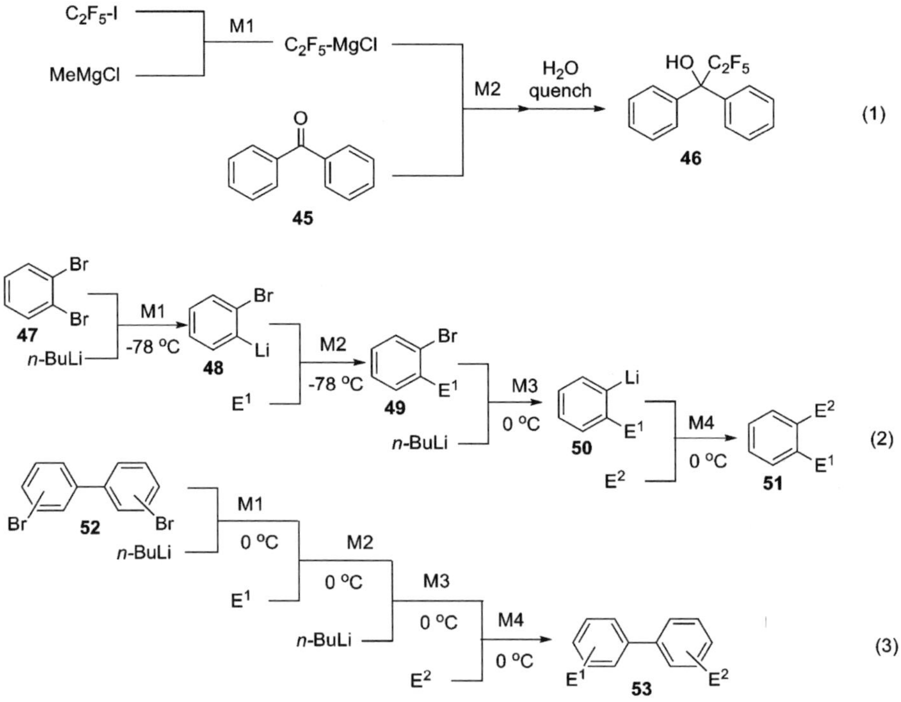

Scheme 4.13 Metal–halogen exchange reactions in microreactors (E = electrophile, M = microreactor).

at $-6\,^{\circ}$C, while the addition to benzophenone required a residence time of 8 min, yielding 86% of product after aqueous work-up.

The lithium–bromide exchange reaction of *o*-dibromobenzene to produce *o*-bromophenyllithium typically requires reaction temperatures of $-110\,^{\circ}$C or below in the batch process. At temperatures above $-110\,^{\circ}$C, LiBr is expelled from *o*-bromophenyllithium to form reactive benzyne. With a shortened residence time in the microreactor, the reaction could be conducted at $-78\,^{\circ}$C, generating *o*-bromophenyllithium that was effectively trapped with various electrophiles.[32] *Ortho*-functionalised benzene derivatives **51** were obtained in good yields starting from *o*-dibromobenzene *via* a continuous four step sequence [Scheme 4.13, eqn (2)]. The initial bromine–lithium exchange, followed by nucleophilic addition to an electrophile, was carried out at $-78\,^{\circ}$C, while the second lithium–bromide exchange/electrophilic quench sequence was carried out at $0\,^{\circ}$C.

This microreactor-based method was taken one step further by application to the selective mono-lithiation of dibromobiaryls [Scheme 4.13, eqn (3)].[33] Dilithiation of the dibromoaryl substrates was substantially reduced using microreactors compared to the batch process, where it is a serious problem. Unsymmetrically substituted biaryls **53** were prepared using this method *via* sequential introduction of *n*-BuLi and electrophiles.

4.4.6.2   Amide Bond Formation

Amide bond formation is a fundamental reaction in medicinal chemistry; amide bonds are featured as key structural elements in a vast number of pharmaceutically interesting compounds.[34] Aluminium-mediated amide bond formation directly *via* the reaction of amines with esters[35] is an attractive alternative to the more commonly applied three-step sequence of ester hydrolysis, activation and treatment with an amine. Trialkylaluminium reagents, however, are highly pyrophoric and the aluminium–amide intermediates are unstable and tend to decompose with strong exotherms.

Recently, a microreactor-based method was developed to resolve these issues.[36] The accumulation of significant amounts of the aluminium–amide is avoided by its continuous formation followed by immediate reaction within the small internal volume of the microreactor. The utility of this methodology was established by application to the synthesis of two pharmaceutically active molecules—rimonabant **56**,[37] a central cannabinoid receptor antagonist, and the radiation therapy enhancing substance efaproxiral **59**[38] (Scheme 4.14).

4.4.6.3   Swern–Moffat Oxidation

The use of microreactors can be beneficial for reactions involving unstable intermediates, which can be reacted immediately before extensive decomposition can occur. The Swern–Moffat oxidation is a common reaction used in synthetic laboratories for the oxidation of primary and secondary alcohols.

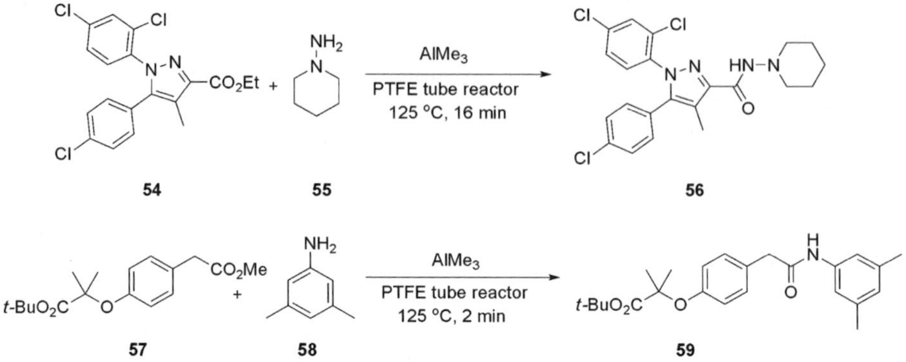

54 **55** **56**

57 **58** **59**

Scheme 4.14 Aluminium-mediated amide bond formation.

Cryogenic conditions are typically required for this oxidation, in particular for the step involving the activation of dimethylsulfoxide (DMSO) with trifluoroacetic anhydride to produce a trifluoroacetoxydimethylsulfonium salt intermediate that is unstable and can undergo an exothermic Pummerer rearrangement at temperatures above –30 °C.[39] Using microflow systems, Swern–Moffat oxidations could be conducted at temperatures between –20 and 20 °C, with good yields of the desired oxidised products.[40]

4.4.6.4 *Ring Expansion Reactions with Diazo Acetate*

Significant safety concerns can arise during the scale-up of reactions involving reagents such diazo compounds and azides. Zhang and co-workers have studied the ring expansion of *N*-Boc-4-piperidone with BF$_3$.Et$_2$O activated ethyl diazoacetate in ether.[41] This reaction presents a number of challenges for scale-up such as the high exothermicity of the addition of BF$_3$.Et$_2$O to the mixture of *N*-Boc-4-piperidone and ethyl diazoacetate. The diazo compound used is thermally unstable and potentially explosive. The reaction profile also shows an initiation period followed by rapid reaction once 60% of the material has been added, resulting in a sudden release of large amounts of nitrogen gas. These risks were effectively minimised owing to the reduced inventory and small internal volume of a microreactor system, which was successfully applied towards the scale-up of this reaction (a throughput of 91 g/h was achieved, with a residence time of 1.8 min) (Scheme 4.15).

4.4.6.5 *Nitration*

Nitration is an industrially important process, with standard process conditions requiring the use of a combination of sulfuric acid and nitric acid as nitrating reagent. Unfortunately, nitration is a highly exothermic and temperature sensitive reaction, making it one of the more difficult reactions to scale up. The small volume and increased heat and mass transfer capability of microreactors

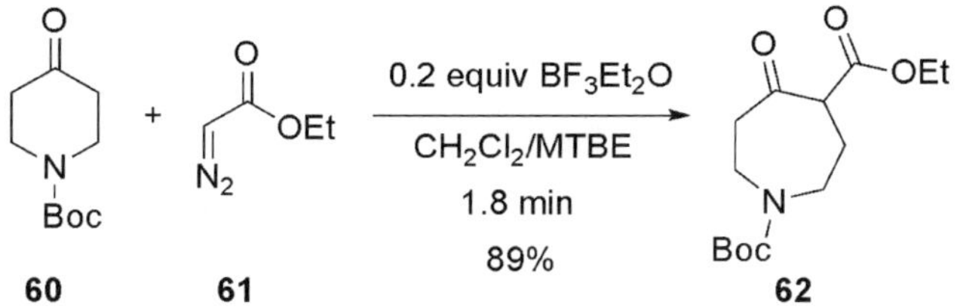

Scheme 4.15   Ring expansion reaction with diazo compound **61**.

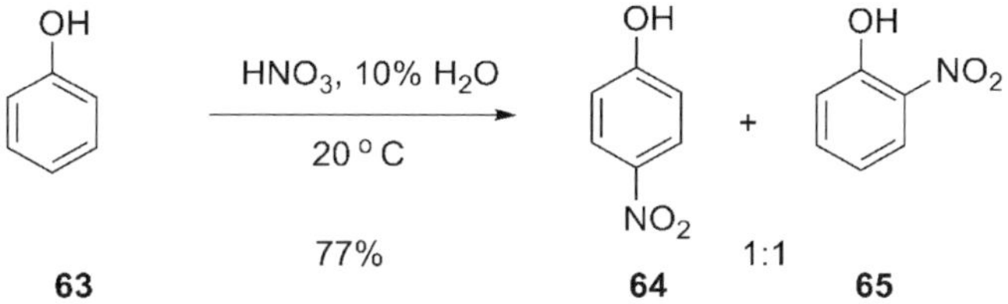

Scheme 4.16   Nitration reactions in a microreactor.

allow for better control of the nitration reaction conditions. Researchers at Lonza studied the nitration of phenol using microreactors with a channel width of 500 μm and an internal volume of 2.0 mL (Scheme 4.16).[42] Screening showed that the nitration was most efficient under neat conditions at 20 °C and without the addition of sulfuric or acetic acid. Under these conditions, autocatalysis started spontaneously within the mixing zone, allowing for better temperature control. Additionally, nitration in the microreactor system was found to suppress polymer formation by a factor of ten compared with the corresponding batch process.

4.4.6.6   Fluorination

Fluorinated organic compounds are gaining prominence amongst pharmaceutically active substrates owing to their unique physical and biochemical properties.[43] The most commonly used fluorinating reagent, diethylaminosulfur trifluoride (DAST),[44] is difficult to handle at a larger scale due to its high reactivity and potential explosive decomposition at temperatures above 90 °C.

Microreactor-based methods have been developed for the deoxyfluorination of various organic substrates. Negi *et al.* studied the deoxyfluorination of a steroid **66** in a perfluoroalkoxy (PFA) tube reactor [Scheme 4.17, eqn (1)][45] using deoxofluor bis(2-methoxyethyl)aminosulfur trioxide instead of DAST owing to its improved safety profile. A detailed kinetic study was carried out to determine the governing reaction parameters.

The fluorination of various alcohols, ketones and carboxylic acids using DAST as the fluorinating agent has been studied.[46] A continuous flow system

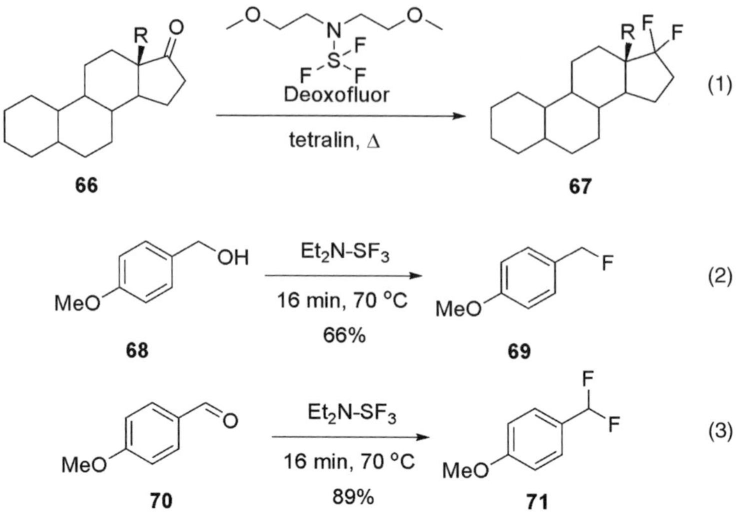

Scheme 4.17 Fluorination.

consisting of a 16 mL tube reactor (PTFE tube), a simple HPLC back-pressure valve and HPLC pumps was used for reactions that were completed in 16 min at 70 °C, using one equivalent of DAST [Scheme 4.17, eqn (2) and eqn (3)]. Ley and co-workers have reported similar fluorination reactions using a microflow reactor coupled with an in-line purification system consisting of a glass column filled with equal amounts of solid calcium carbonate and silica gel.[47]

4.4.7 Biphasic Liquid–Liquid Reactions

The application of biphasic liquid–liquid reactions to microfluidic systems is particularly interesting because of the opportunities to manipulate flow patterns between the two immiscible phases.[48] Improved mass transfer and mixing between the two phases can be achieved, potentially leading to significant enhancements in reaction rates and selectivities. The separation of reagents and reaction products can often be simplified in a biphasic system.

An early example of a biphasic liquid–liquid reaction in a microreactor was reported focusing on the isomerisation of allylic alcohols to ketones using various metal catalyst and water soluble ligands in different aqueous/hydrocarbon solvent systems (Scheme 4.18).[49] The various catalysts and substrates were introduced simultaneously by pulsed injections into the micromixer to generate emulsion droplets which were then carried through the integrated microchannel tube.

Mukaiyama aldol reactions in a biphasic fluorous/organic solvent system were described by Mikami *et al.*[50] A very dilute solution of the lanthanide amide catalyst $Sc[N(SO_2C_8F_{17})_2]_3$ in $CF_3C_6F_{11}$ was introduced as the fluorous phase, with the organic phase consisting of a solution of benzaldehyde and the

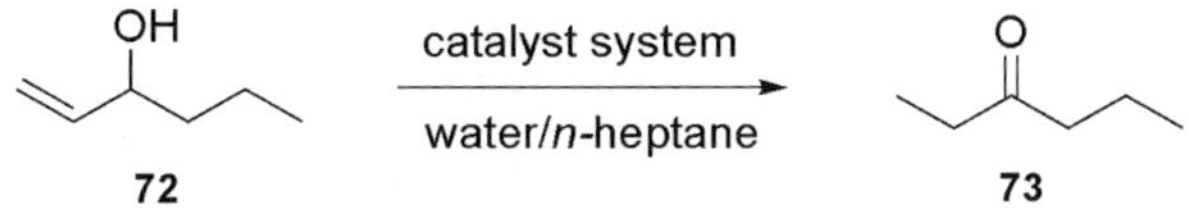

Scheme 4.18 Metal-catalysed isomerisation of allylic alcohol.

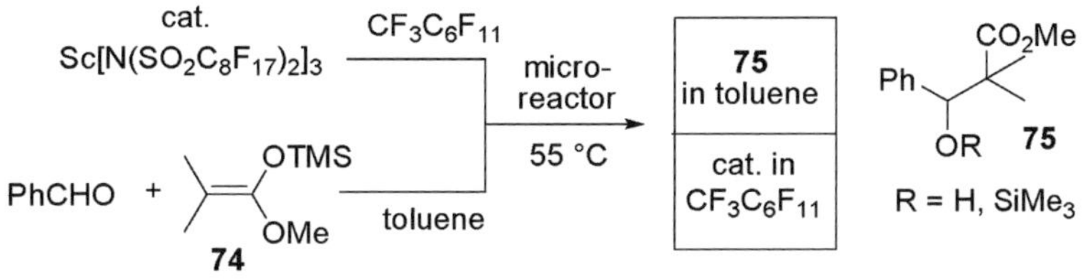

Scheme 4.19 Mukaiyama aldol reaction.

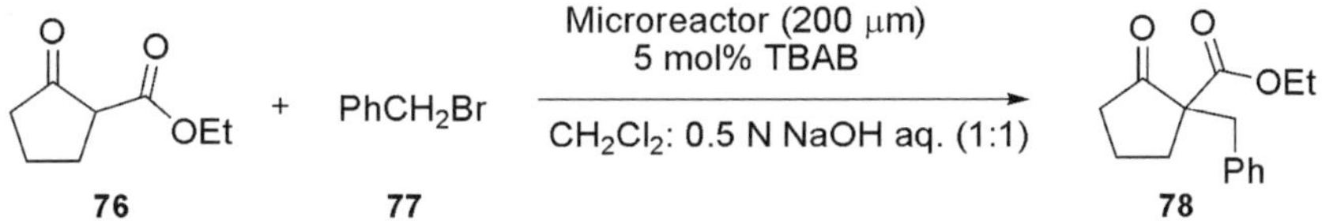

Scheme 4.20 Phase transfer alkylation reaction.

trimethylsilyl enol ether of methyl 2-methylpropanoate in toluene (Scheme 4.19). The reactions were completed within seconds even with <0.1 mol% of catalyst, which remained in the fluorohydrocarbon phase and was recoverable. Baeyer–Villiger oxidations catalysed by $Sc[N(SO_2C_8F_{17})_2]_3$ were also investigated using a similar biphasic system; reaction rates and lactone selectivities were much higher compared to batch results.[51]

Phase transfer alkylation reactions using tetrabutylammonium bromide (TBAB)[52] as catalyst have also been studied using a microreactor system. Alkylation reactions of ß-ketoesters were found to be faster than in the analogous batch process possibly due to an increased interfacial area (Scheme 4.20). The effect of width of the microchannels in the reactor was also examined. Whilst, in all cases, a segmented type of flow (organic phase droplets within the aqueous flow) was observed, the narrower channels resulted in faster reaction rates, suggesting that interfacial contact is increased with smaller droplets. The phase transfer alkylation of a malonic ester with catalytic tetrabutylammonium hydrogen sulfate was also investigated.[53]

Wirth and co-workers reported a tandem diazotisation–Heck reaction performed in a microreactor device using segmented flow (Scheme 4.21). The two different liquid phases were injected *via* a T-junction into the reactor resulting in serial separation within the channel.[54] The yields of the Heck products were

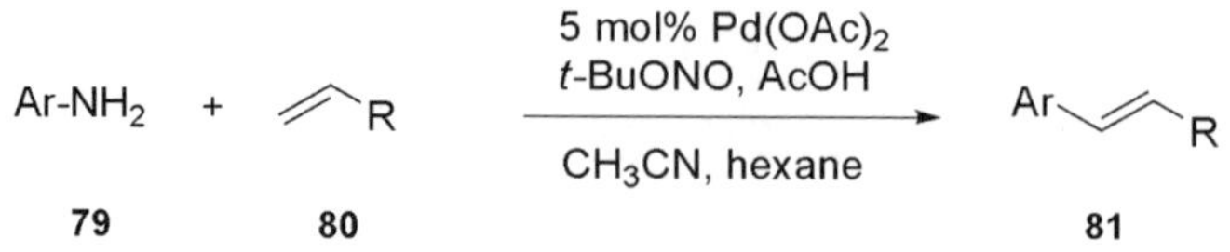

$$Ar\text{-}NH_2 \ + \ \diagup\!\!\diagup R \ \xrightarrow[\text{CH}_3\text{CN, hexane}]{\substack{5 \text{ mol\% Pd(OAc)}_2 \\ t\text{-BuONO, AcOH}}} \ Ar\diagdown\!\!\diagup\!\!\diagdown R$$

79   **80**   **81**

Scheme 4.21   Heck reaction in a microreactor using segmented flow.

observed to be slightly higher when segmented flow was used compared to the reactions conducted in single flow.

4.5   Liquid–Gas Reactions

Synthetic chemical transformations using toxic and/or corrosive gases are in general challenging to perform due to the hazardous and strongly reactive nature of the gaseous reactants. A further limiting factor in the success of carrying out gas–liquid transformations is mass transport between the two phases, which is improved in microfluidic devices due to increased interface in contact. Specially designed liquid–gas microreactors[55] allow the careful control of gas flow and pressure, and ensure the uniform distribution and contact between the gas and liquid phases. Gas–liquid separators can also be integrated to separate the gaseous phase after completion of the reaction.[3]

The general applicability of continuous flow microreactors towards the area of liquid–gas transformations has been demonstrated for fluorinations[56] and chlorinations[57] (with elemental halogen), nitrations[58] and oxygenations.[59] Homogeneous palladium-catalysed gas–liquid reactions such as aminocarbonylations,[60] carbonylative Sonogashira couplings,[61] Stille reactions,[62] microwave-assisted continuous-flow Suzuki reactions[63] and Heck reactions[64] have also been developed under microflow conditions.

4.5.1   Oxidation with Ozone

The oxidation of intermediates in batch reactions using ozone gas can be problematic due to the high reactivity and toxicity of ozone. By using a specially designed silicon–Pyrex microreactor, the ozonolytic cleavage of olefins was performed at room temperature instead of the established batch temperatures of –78 °C (Scheme 4.22).[59c] Olefins, as well as phosphite esters and primary amines, were oxidised to generate the corresponding oxidation products with extremely fast reaction times (0.3 s) and in excellent yields, as determined by gas chromatography (GC) analysis.[59c]

4.5.2   Singlet Oxygen Oxidation

The generation of singlet oxygen and its application to large-scale synthesis can be dangerous due to the explosion risks associated with highly oxygenated

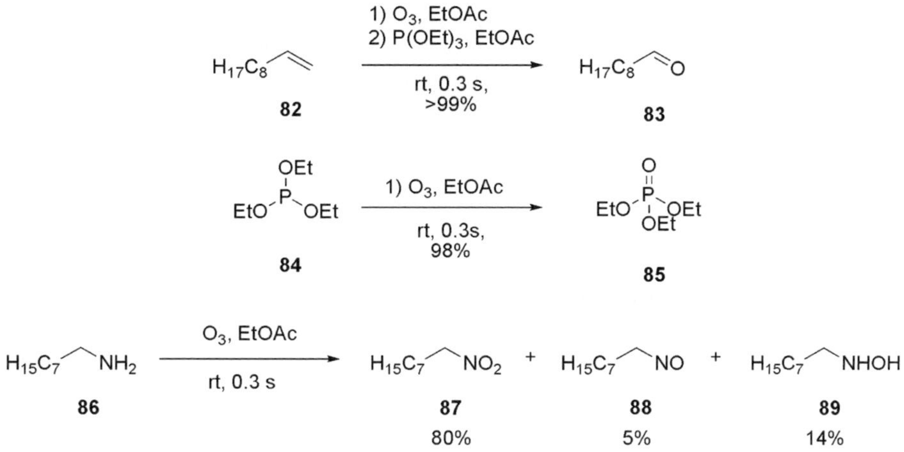

Scheme 4.22 Oxidations using ozone performed in a silicon–Pyrex microreactor.

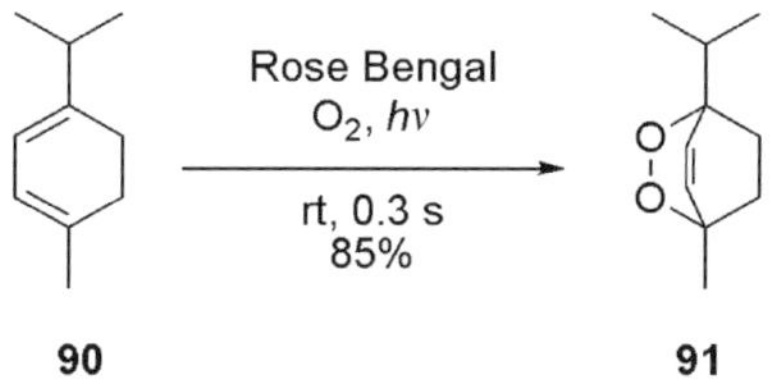

Scheme 4.23 Singlet oxygen addition to α-terpinene **90**.

flammable organic liquids. De Mello and co-workers investigated the cyclo-addition of singlet oxygen to α-terpinene using a nanoscale glass micro-reactor.[59b] The above-mentioned safety risks were avoided in the microreactor as only miniscule amounts of oxygenated solvent existed at a given time period. Factors relating to the inefficient irradiation of the photosensitiser required for the photochemical generation of singlet oxygen were also not an issue under these conditions (see Section 4.4.1 for a discussion on the advantages and uses of microreactors for photochemical reactions). α-Terpinene **90** was continuously oxidised by on-chip generated singlet oxygen, producing the natural product ascaridole **91** in a GC conversion of 85% in less than 5 s reaction time (Scheme 4.23).[59b]

4.5.3 Fluorination

Fluorinations using elemental fluorine are highly exothermic and are difficult to control using conventional equipment. The rapid formation of HF by contact with water or moisture renders the choice of appropriate equipment even more difficult. By taking advantage of the efficient mass and heat transport in

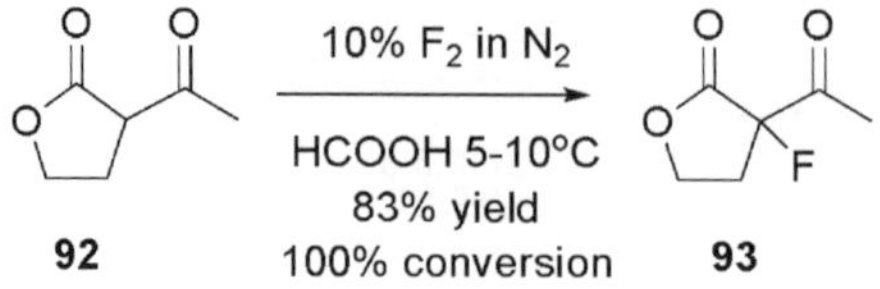

Scheme 4.24   Monofluorination of 1,3-dicarbonyls.

microfluidic devices, direct fluorination of 1,3-ketoesters and 1,3-diketones at
−8 to 10 °C was carried out in a nine-channel microreactor. The selective
monofluorination of 1,3-dicarbonyl products with fluorine (as a 10% (v/v)
mixture with nitrogen) was achieved successfully with high conversions and
reasonable yields (Scheme 4.24).[56d]

4.5.4   Chlorination

The photochemical chlorination of toluene-2,4-diisocyanate **94**[57b] was investigated in a nickel-coated, Institut für Mikrotechnik Mainz (IMM) falling film
reactor[65] (Scheme 4.25). The top plate of the falling film reactor consisted
partially of a quartz slide to allow for controlled irradiation of the reaction
mixture to generate free chlorine radicals. Channel dimensions of 300 μm depth
allowed for efficient penetration of light into the reaction flow. The formation
of by-products could be kept at a minimum with a low local concentration of
active chlorine radicals. Chlorination at the benzylic position of **94** proceeded
with 55% conversion to give **95** with 80% selectivity, whereby the formation of
undesired side product **96** was suppressed. In contrast, the batch reaction gave
higher conversions (65%) but decreased selectivity (45%).[57b]

4.5.5   Cross-coupling Reactions

Carbonylative cross-coupling reactions of aryl halides and primary amines are
useful synthetic transformations for the formation of secondary amides.
However, lengthy reaction times and the high pressures of toxic carbon monoxide generally required can render these reactions difficult to perform. The
application of microreactor technology to these types of reactions would thus
lead to overall improvements in the process safety.

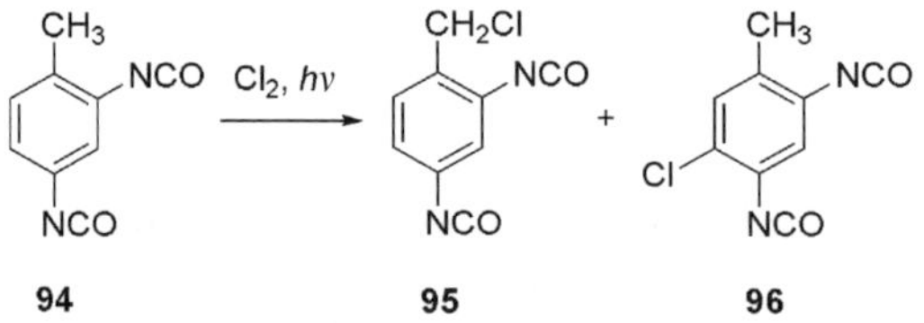

	Conversion	Selectivity
Batch	65%	45%
Microreactor	55%	80%

Scheme 4.25   Direct chlorination of toluene-2,4-diisocyanate **94**.

Aminocarbonylation reactions based on a Pd(OAc)$_2$-Xantphos catalyst system have been carried out in silicon-glass microreactors (Scheme 4.26).[60b] In contrast to conventional bench-top reactions at near atmospheric pressure, the aminocarbonylation of aryl halides **97** in the microreactor produced significant amounts of ketoamides **100** in addition to amides **99**.

The formation of the ketoamide adducts was attributed to the greatly improved interfacial contact between the gas and liquid phases, and efficient mixing in the miniaturised channel systems. By variation of the reaction conditions (*e.g.* reaction temperature, carbon monoxide pressure) for electron-rich as well as electron-deficient aryl halides, the relative ratios of amides and α-ketoamides were manipulated towards excellent selectivities.[60b]

Homogeneous palladium-catalysed carbonylative Sonogashira cross-coupling between iodobenzene **102** and phenylacetylene **103** in an ionic liquid, [bmim]PF$_6$, were reported by Ryu *et al.* (Scheme 4.27).[61] Performing the

Aryl Halide	P (bar)	T (°C)	time (min)	conversion (%)	Amide (%)	Ketoamide (%)
MeO—⬡—I	7.9	146	3.3	100	68	28
	7.9	116	4.2	100	35	65
NC—⬡—Br	2.7	160	7.1	100	83	0
	14.8	109	6.6	99	32	57

Scheme 4.26 Aminocarbonylations performed in a silicon–glass microreactor.

	P (atm)	104 (%)	105 (%)
Microflow System (12 min)	5	83	0
	3	80	0
Autoclave reactor (60 min)	5	25	60
	3	14	65

Scheme 4.27 Carbonylative coupling of iodobenzene and phenylacetylene in a stainless steel microreactor.

reaction in a stainless steel microreactor system clearly demonstrated the advantages of microreactor technology over conventional procedures due to the highly efficient mixing of the different phases. In contrast to the reactions performed in traditional autoclaves, the microflow system yielded the desired carbonylation product **104** selectively with significantly shortened reaction times.

4.6   Liquid–Gas–Solid Reactions

Triphasic catalytic reactions such as hydrogenations,[66] carbonylations,[67] or oxidations[68] are important synthetic transformations. As with the liquid–gas reaction systems described in the previous section, these multiphasic transformations suffer from long reaction times due to the poor mixing of reactants and small interfacial areas between the different phases. Continuous flow microfluidic devices have been developed[55] such that triphasic reactions can be carried out safely, with reasonable reaction times and improved selectivities.

4.6.1   Hydrogenation

An efficient catalytic hydrogenation system based on a glass microreactor with covalently bound microencapsulated palladium has been developed for the rapid reduction of alkenes, alkynes and benzyl groups (Scheme 4.28a).[66d] Notably, alkyne **110** was selectively reduced in the presence of the benzylic protection group.[66d] The solubility of hydrogen in the organic solvent used for the hydrogenation is often a key factor in determining reaction rates. In order to improve the existing microreactor protocol, Kobayashi and co-workers further investigated these reactions in supercritical carbon dioxide (scCO$_2$) (Scheme 4.28b).[66e] ScCO$_2$ attracts much interest as a replacement for traditional solvents in synthetic

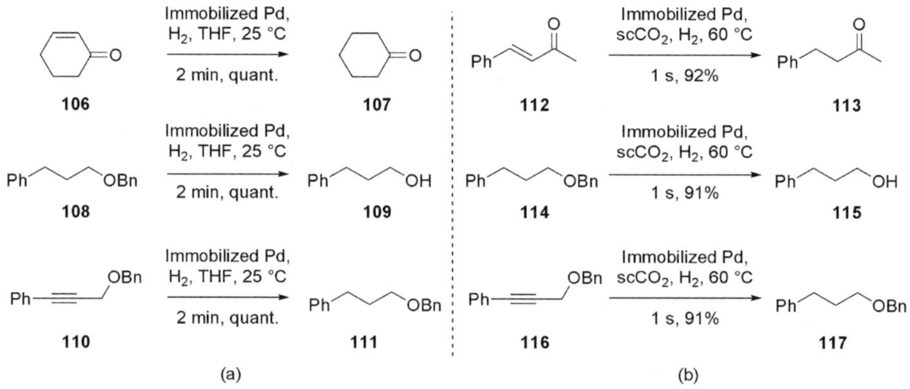

(a)

106 → **107**: Immobilized Pd, H$_2$, THF, 25 °C; 2 min, quant.

108 (Ph~~~OBn) → **109** (Ph~~~OH): Immobilized Pd, H$_2$, THF, 25 °C; 2 min, quant.

110 (Ph—≡—OBn) → **111** (Ph~~~OBn): Immobilized Pd, H$_2$, THF, 25 °C; 2 min, quant.

(b)

112 (Ph~~~C(O)CH$_3$) → **113** (Ph~~~C(O)CH$_3$): Immobilized Pd, scCO$_2$, H$_2$, 60 °C; 1 s, 92%

114 (Ph~~~OBn) → **115** (Ph~~~OH): Immobilized Pd, scCO$_2$, H$_2$, 60 °C; 1 s, 91%

116 (Ph—≡—OBn) → **117** (Ph~~~OBn): Immobilized Pd, scCO$_2$, H$_2$, 60 °C; 1 s, 91%

Scheme 4.28   Hydrogenation reactions carried out in a palladium wall coated microreactor. (a) Reaction carried out in THF. (b) Reactions carried out in supercritical CO$_2$.

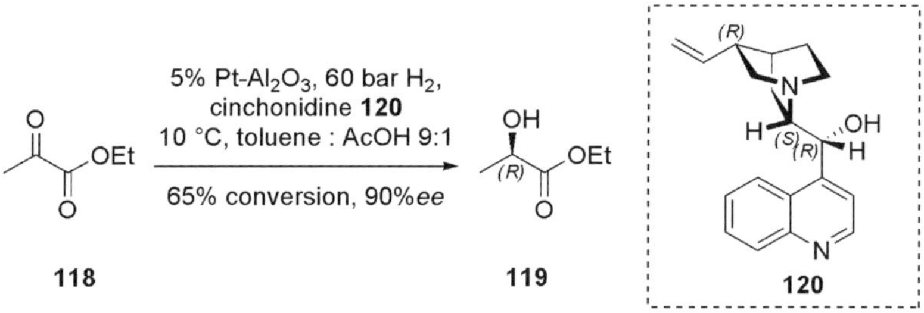

Scheme 4.29 Enantioselective hydrogenation in continuous flow.

organic chemistry due to its non-toxicity, low costs, and its high diffusivity and miscibility with gases.[69] Consequently, reactions can proceed without the mass transport limitations often encountered in traditional multiphase systems.

The importance of enantioselective hydrogenation for the synthesis of optically pure building blocks or targets is well-known in the scientific community. α-Hydroxy esters belong to a class of synthetically useful intermediates that can be readily accessed using enantioselective heterogeneous catalytic hydrogenation.[70] This reaction was intensively investigated in continuous-flow,[71] and recently a protocol using the commercially available hydrogenation reactor H-Cube®, an apparatus that combines continuous flow microchemistry with *in situ* hydrogen generation by the electrolysis of water,[72] was reported (Scheme 4.29).[66j] A solution of α-ketoester **118** was flowed through a prepacked catalyst cartridge (5% palladium on alumina) with 60 bar of hydrogen pressure. Using cinchonidine **119** as an enantiopure additive, (*R*)-ethyl lactate **120** was obtained with 90% enantiomeric excess (ee).[66j]

4.6.2 Reductive Amination

Reductive aminations are key transformations in the synthesis of many pharmaceutically active intermediates. Nevertheless, some of the traditional protocols are often hampered by the reversible nature of the reaction, functional group incompatibility and over-reduction. The reduction of C-aryl imines is especially prone to over-reduction, producing secondary amine products highly contaminated with primary amines. Aromatic imines have been reduced with high chemoselectivities using a H-Cube® hydrogenation reactor (Scheme 4.30).[66f] Hydrogenation of the imines **121** and **123** using a prepacked catalyst cartridge (10% palladium on charcoal) and a hydrogen pressure of 20 bar provided the desired amines **122** and **124** quantitatively. Notably, other readily reducible functional groups such as nitrile and pyridine were not affected.[66f]

4.6.3 Aminocarbonylation

Carbon monoxide insertion is a convenient synthetic method for accessing various common functional groups such as amides and esters (see Section 4.5.5

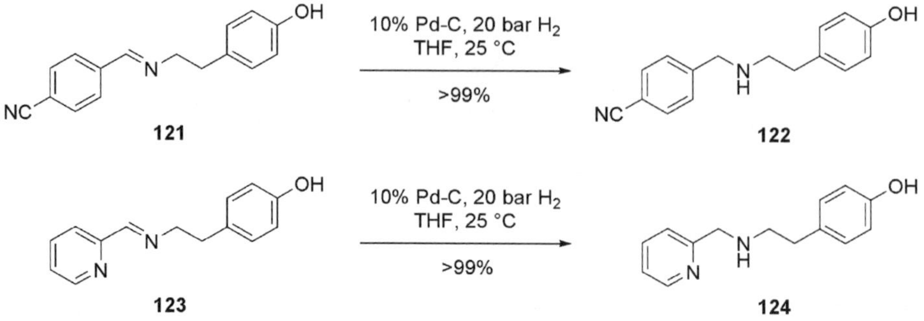

Scheme 4.30   Selective hydrogenation of C-aryl imines **121** and **123** using the H-Cube®.

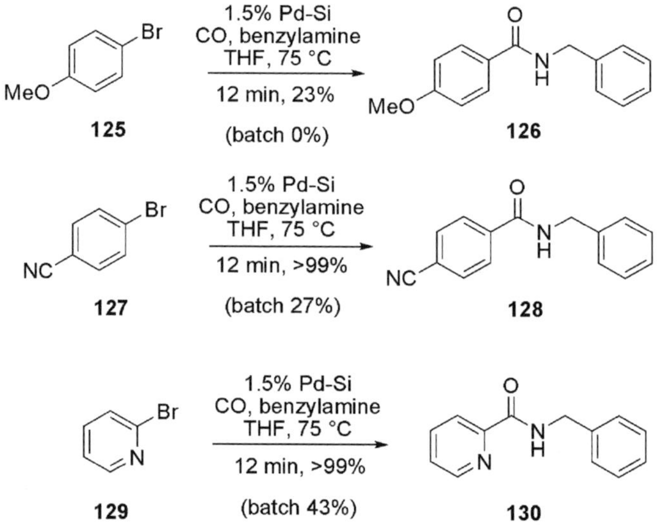

Scheme 4.31   Aminocarbonylation in catalyst-packed Teflon® tubing.

for related examples). The reaction rate depends strongly on the nature of the catalyst, the aryl halide and the nucleophile coupling partners, and on reaction parameters such as temperature and carbon monoxide pressure. Microreactors offer the potential for careful reaction time and temperature control under high pressures, permitting an increased ratio of starting material and active catalyst to carbon monoxide.

Long *et al.* reported the aminocarbonylation of aryl halides **125**, **127**, and **129** in continuous-flow, using Teflon tubing packed with a silica-supported palladium catalyst (Scheme 4.31).[67a] The carbonylations were completed within 12 min at 75 °C with much higher (GC) yields compared to the analogous reaction run in conventional glassware under the same conditions.

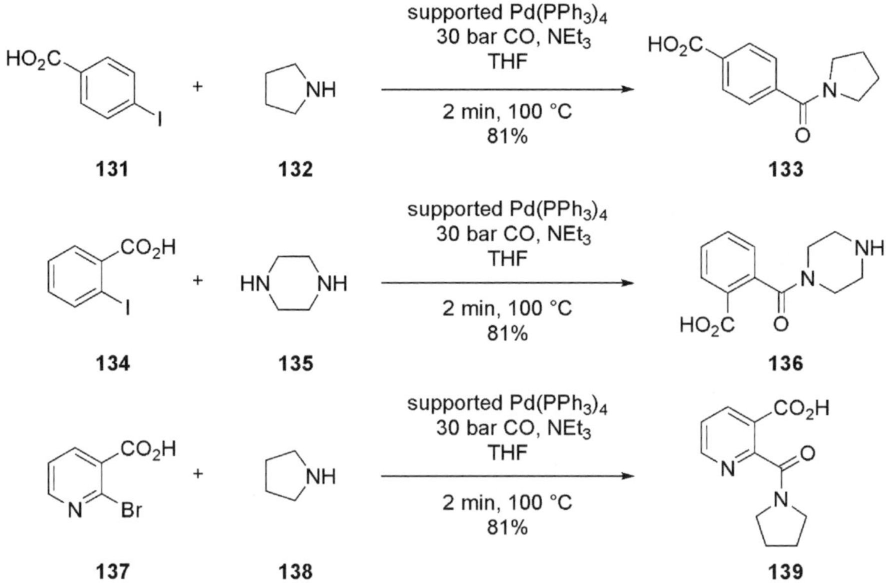

Scheme 4.32 Synthesis of dicarboxylic acid derivatives by aminocarbonylation in the X-Cube[®].

Selectively derivatised dicarboxylic acid monoamides, commonly occurring subunits in biologically active compounds, were synthesised using the commercially available X-Cube[®72] (Scheme 4.32).[67b] A prepacked catalyst cartridge [polymer supported Pd(PPh$_3$)$_4$] and 30 bar CO pressure were applied to yield monoamides **133**, **136** and **139** in 2 min. Remarkably, no second aminocarbonylation occurred with piperazine as the amine nucleophile.

4.6.4 Alcohol Oxidation

The metal-catalysed oxidation of alcohols to aldehydes and ketones is an important chemical transformation for the synthesis of building blocks and intermediates. The use of molecular oxygen as an oxidant is limited due to catalyst deactivation and aggregation upon the formation of palladium black, as well as the safety hazards associated with large quantities of highly flammable organic solvent–oxygen mixtures. Supercritical carbon dioxide (scCO$_2$) on the other hand is benign, non-flammable and highly miscible with molecular oxygen.

The triphasic oxidation of primary and secondary alcohols to their corresponding carbonyl compounds using microstructured continuous-flow devices has been studied by Baiker *et al.*[68c] A tubular fixed bed reactor (0.5% palladium on alumina) was used with scCO$_2$ as solvent for the oxidation of 1-octanol **140**. 1-Octanal was formed with 73% selectivity over oxidation products **142** and **143**, albeit with a rather low conversion of 3.3%. Increases in

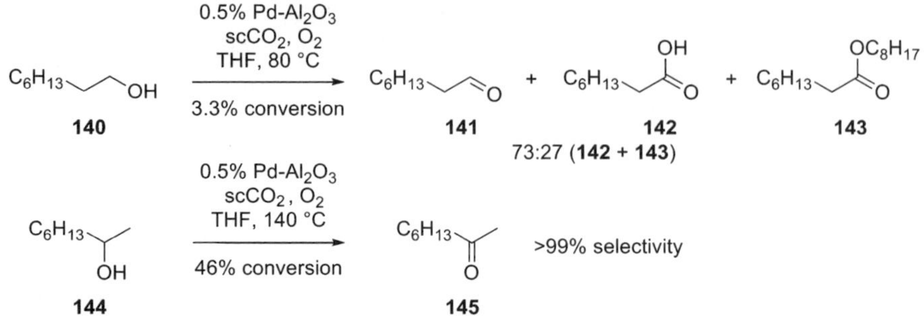

Scheme 4.33   Oxidation using molecular oxygen performed in continuous flow.

temperature and pressure as well as oxygen concentration resulted only in decreased selectivities (Scheme 4.33). The oxidation of 2-octanol **144** proceeded to give the corresponding ketone with moderate conversion, and excellent selectivity.[68c]

4.7   Solid Supports and Monolith-bound Reagents in Continuous Flow

The use of free-flowing insoluble reagents in microflow systems is somewhat limited due to the high propensity and likelihood of build-up and subsequent flow blockage. As may be apparent from the examples discussed in Section 4.6, solid reagents used in flow chemistry are for the most part immobilised, either directly to the reactor walls or *via* a supporting surface. Supported reagents and scavengers have impact at the way chemistry is performed.[73]

Key advantages associated with supported reagent synthesis include:

- faster and simplified procedures for the generation of a large number of compounds;
- ease of purification of intermediates and products;
- the possibility of using excess reagents without complicating post-reaction purification.

After many successful applications of solid phase synthesis in the batch mode, the next logical step was to combine the use of solid supported reagents and scavengers with continuous flow technology.

4.7.1   Solid-supported Reagents

An early example of solid supported reagents in flow was demonstrated for the synthesis of the natural product grossamide.[74] Ferulic acid **146** was effectively coupled to amine **147** using solid-supported 1-hydroxybenzotriazole (HOBt) in combination with bromo-tris-pyrrolidino phosphoniumhexafluorophosphate

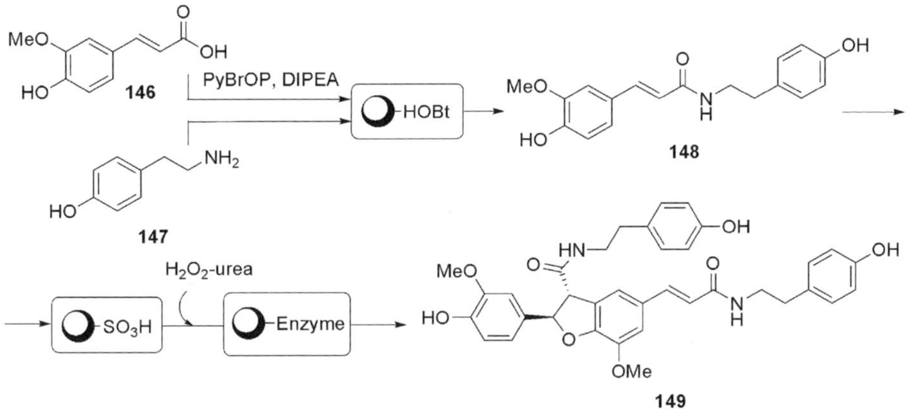

Scheme 4.34 Polymer supported continuous flow synthesis of grossamide (Enzyme = horseradish peroxidase type II; DIPEA = di(isopropylethylamine)).

(PyBrOP) (Scheme 4.34). The resulting reaction stream was flowed through an amine-scavenging column, then treated with a solution of hydrogen peroxide–urea complex and converted directly to grossamide **149** using a polymer supported horseradish peroxidase type II enzyme. This short synthesis elegantly demonstrated the concept and power of using solid supported reagents in continuous flow synthesis.

This concept was further applied to the synthesis of short peptides using Boc-(*tert*-butyloxycarbonyl), Cbz- (carbobenzyloxy), or Fmoc-(fluorenyl methoxy carbonyl) protected amino acids.[75] Amino acids were treated with either PyBrOP [Scheme 4.35, eqn (1)], or converted into their corresponding anhydrides using 2-isobutoxy-1-isobutoxycarbonyl-1,2-dihydroquinoline (IIDQ) before treatment with polymer-supported HOBt [Scheme 4.35, eqn (2)]. The resulting polymer-bound activated ester was directly reacted with another O-protected amino acid. An acidic scavenging resin was used to remove unreacted amine and the reaction stream was concentrated for isolation of the dipeptide. Tripeptides were also synthesised from Cbz-protected amino acids using an iterative version of this methodology, with an added step of flow hydrogenolysis to remove the Cbz- group.[75]

A method for the continuous flow synthesis of various oxazoles using polymer-supported reagents has also been described.[76] Acyl chlorides and isocyanates bearing electron-withdrawing groups were combined in a micro-flow device to form a reaction stream that was passed through a column of polymer-supported 2-*tert*-butylimino-2-diethylamino-1,3-dimethylperhydro-1,3,2-diazaphosphorine (BEMP), resulting in the formation of 4,5-disubstituted oxazoles. An amine-functionalised resin was included one step further down-stream to scavenge unreacted starting materials and improve oxazole purity (Scheme 4.36). Similar reactions were performed with isocyanates substituted with tosyl or phosphonate groups in place of the ester, resulting in the preparation of a focused library of oxazoles.

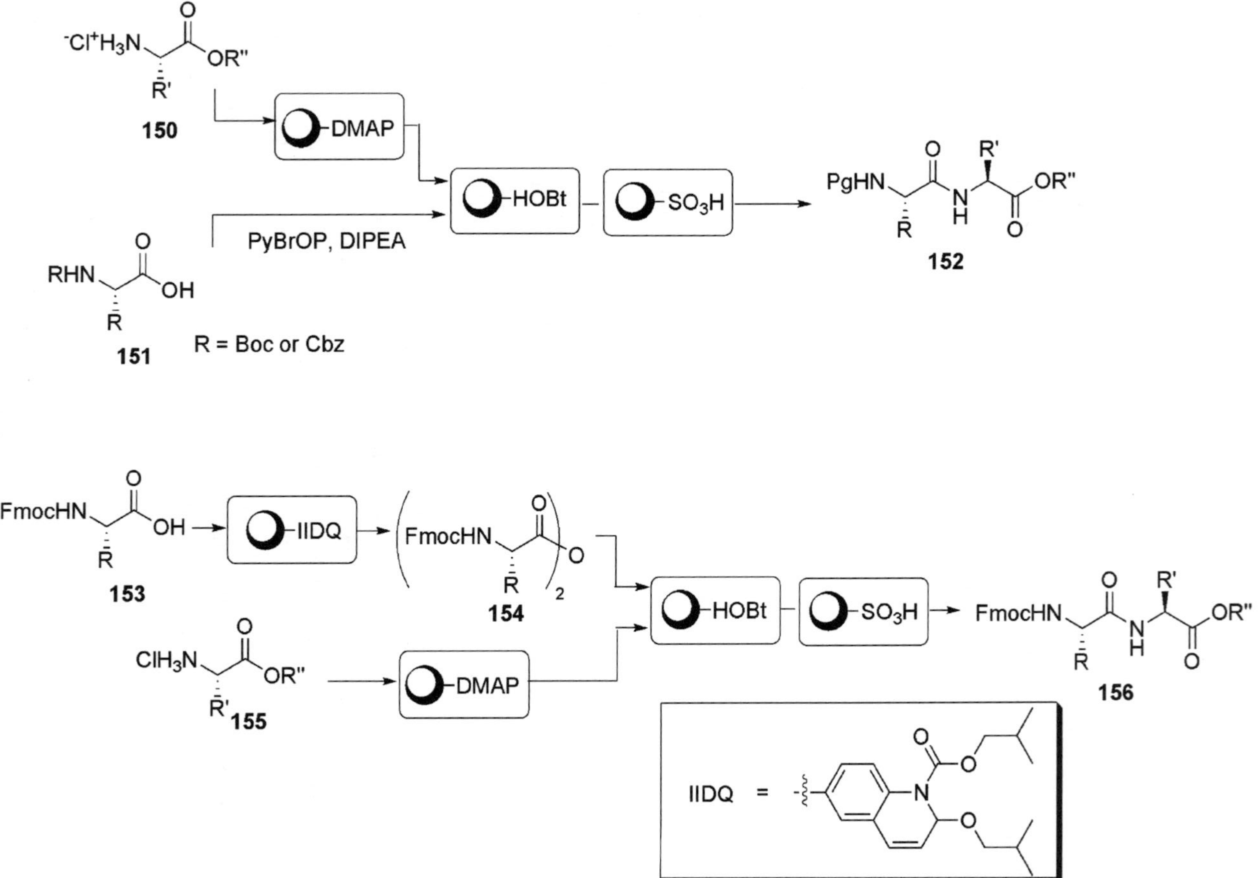

Scheme 4.35   Continuous flow peptide synthesis (DMAP = *N,N*-dimethylamino pyridine).

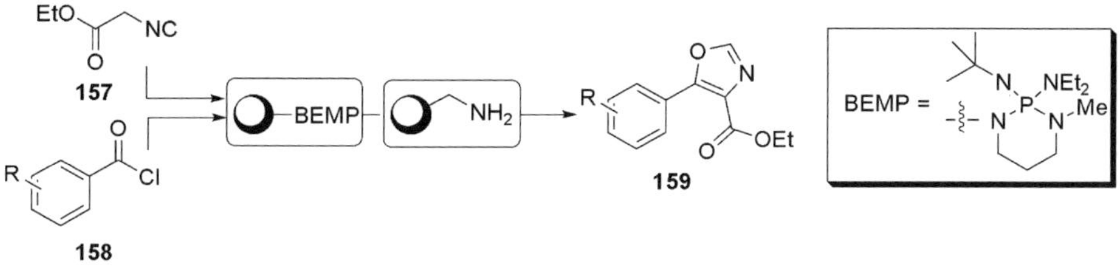

Scheme 4.36 Continuous flow synthesis of oxazoles.

A multistep preparation in continuous flow of the natural product oxomaritidine[77a] **168** was developed on the basis of a previous synthesis[77b] in batch using solid-supported reagents (Scheme 4.37). The seven-step sequence started with the azide substitution of alkyl bromide **162** and perruthenate oxidation of primary alcohol **160** to the corresponding aldehyde **161**. These reactions were performed simultaneously, such that the resulting two streams were immediately combined and reacted *via* an aza-Staudinger reaction by passage through a column of solid-supported dibutylphosphine. The intermediate imine solution was then reduced using continuous flow hydrogenation (H-Cube®) before solvent exchange—from tetrahydrofuran (THF) to DCM—the first and only product handling operation outside the reactor system. The resulting amine solution was trifluoroacetylated in a microreactor, then directed through an acid scavenger column. Oxidative phenolic coupling of the exiting flow stream provided the tricyclic intermediate **167**, and then flow through into a column packed with solid supported base induced cleavage of the trifluoroacetamide protecting group. This resulted in the conjugate addition of the free amine to the 1,2-unsaturated ketone, producing racemic oxomaritidine **168**. The continuous sequence proceeded with 40% overall yield and provided product of 90% purity. Further investigations showed that the oxidative phenolic coupling alone gave a moderate 50% yield and was responsible for the only isolated side product of the seven-step sequence. This impressive synthesis is hitherto the most complex synthesis performed in sequential continuous flow steps, demonstrating that it is possible to reduce the synthetic effort significantly.[77b]

Azide reagents are featured in a number of continuous flow based reactions. For example, syntheses of building blocks important to medicinal chemistry such as 1,2,3-triazoles[78] and amides[79] have been accomplished using azides and polymer-supported reagents. Triazoles were synthesised *via* the copper-catalysed [3 + 2] cycloaddition of alkynes with azides (Scheme 4.38). Copper (I) iodide was attached to an Amberlyst A-21 resin and packed in heated columns that were flushed with azide and alkyne substrates. The scavenging of copper leached into the solution quenching excess azide was performed sequentially using a thio-urea resin and a phenylphosphine resin respectively, delivering products of >95% purity. Undesired side reactions such as Glaser homocoupling was avoided by easy exclusion of oxygen from the flow system.

Microflow reactor technology has also been applied, together with a tagged reagent strategy, for the synthesis of various secondary amides and substituted

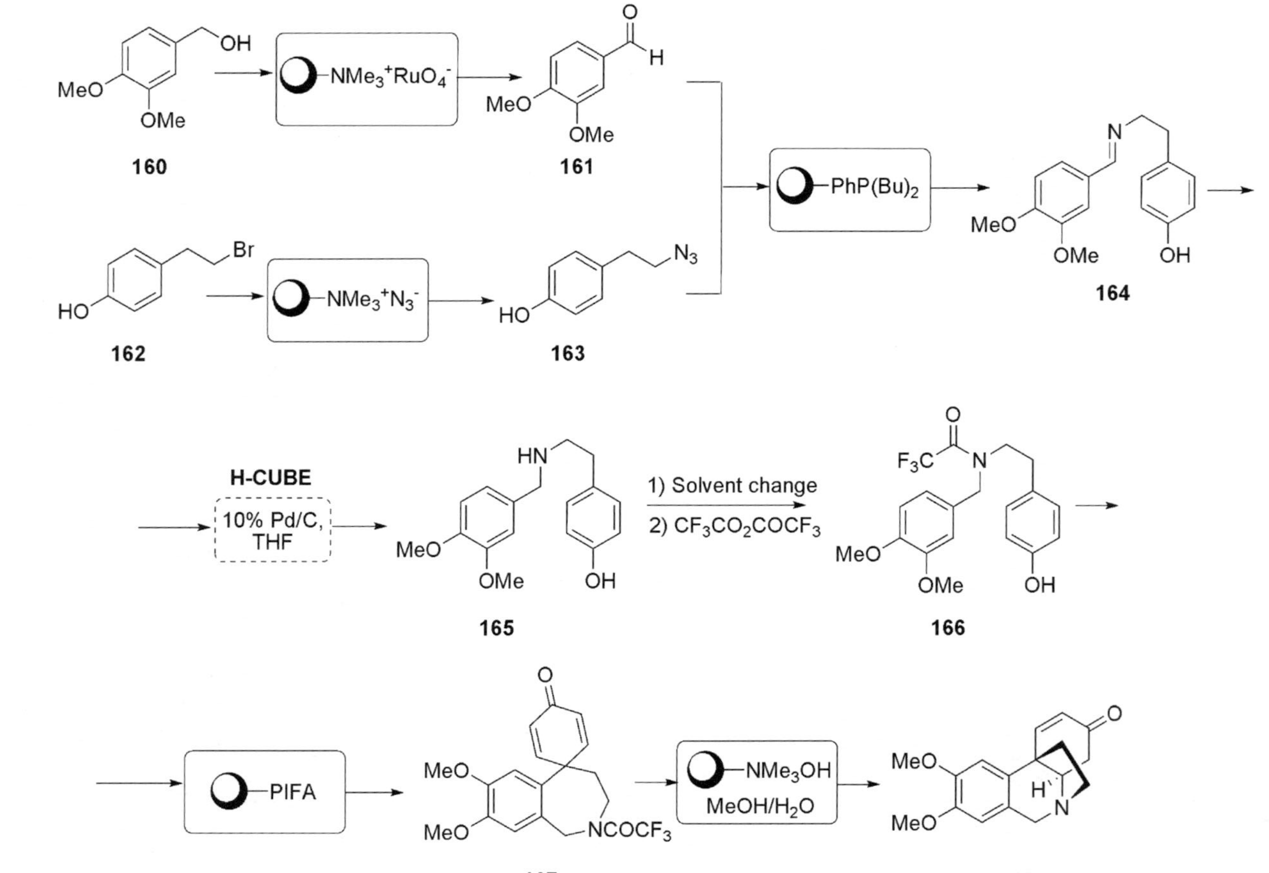

Scheme 4.37 Continuous flow synthesis of (±)-oxomaritidine [PIFA = (ditrifluoroacetoxyiodo)benzene].

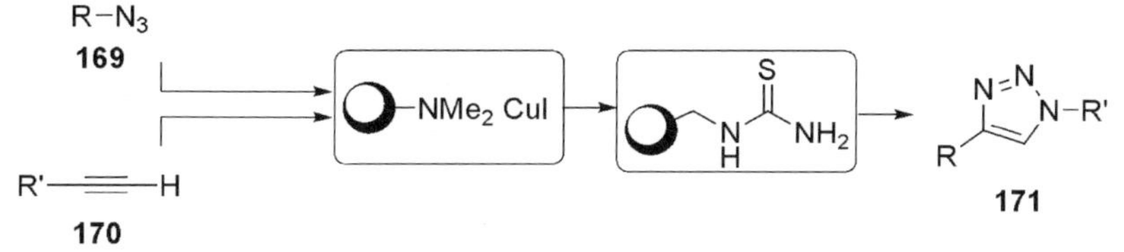

Scheme 4.38 [3 + 2] Cycloaddition of azides and alkynes.

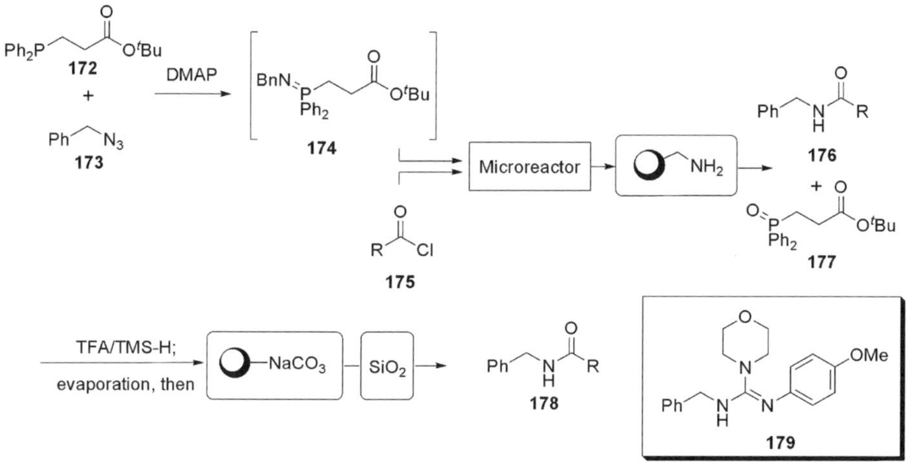

Scheme 4.39 Tagged phosphine reagents to simplify work-up and purification (TMS-H = trimethylsilyl hydride).

guanidines.[79a] Iminophosphorane **174** was generated from the *tert*-butyl ester-tagged phosphine reagent **172** and azide reagent **173**, then directed through a microdevice together with a stream of acyl chloride to form amide **176** and the tagged by-product phosphine oxide **177**. Passage through a benzylamine resin then removed excess acid chloride. Treatment with trifluoroacetic acid (TFA) unmasked the ester tag of the phosphine oxide to the corresponding carboxylic acid, which was removed from the mixture by streaming through a column of solid-supported carbonate base, yielding the desired amide in >95% purity after filtration through a silica plug (Scheme 4.39). A similar strategy was used for preparing trisubstituted guanidines; **174** was reacted with isothiocyanate first in the microreactor, and excess **174** was then scavenged by flow through immobilised isocyanate. The desired guanidine product was formed by input of a solution of secondary amine and the tagged phosphine oxide product was removed in the same process as described for the synthesis of amides.

The Curtius rearrangement is useful for the preparation of amines or their corresponding carbamate from carboxylic acids. However, unstable intermediates such as acyl azides are formed, making the reaction much less

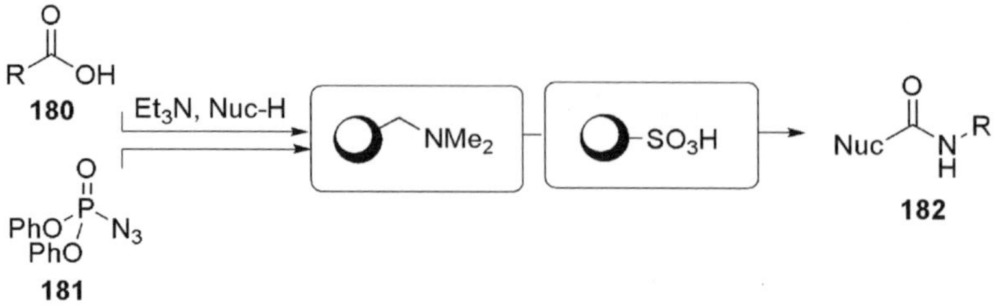

Scheme 4.40 Curtius rearrangement reactions in continuous flow using diphenyl-phosphoryl azide (DPPA).

attractive on a large scale. By utilising continuous flow conditions, the heating of large amounts of sensitive azide reagents and accumulation of acyl azide can be avoided (Scheme 4.40).[79c] A pre-mixed solution of carboxylic acid, triethylamine and the nucleophile was treated with diphenylphosphoryl azide (DPPA) (**181**) and heated to form the isocyanate that was subsequently trapped *in situ* by the nucleophile. A base-scavenging column served to trap unreacted carboxylic acid as well as the newly formed diphenylphosphonic acid. Further scrubbing was done to remove excess triethylamine by inclusion of an acidic resin. In this manner, a series of carbamates as well as ureas were synthesised. An alternative adaptation of the Curtius rearrangement to flow synthesis was also developed based on the use of a monolith-supported azide reagent.[79c]

4.7.2 Solid-supported Catalysts

Supported catalysis in flow is particularly attractive in that not only can the catalysts be removed from the reaction stream, but also that they are recoverable and can be reused. Catalyst loading is less important when compared to homogenous catalysis. Overall catalyst concentrations can consequently be quite high, enabling faster conversion times.

Considering the chemical space that has been made available since transition metals became common in organic chemistry, it is no wonder that continuous flow reactors have frequently been used for this type of chemistry. In an early example, Kirschning and co-workers described the development of a monolithically supported palladium catalyst that proved to be effective for various cross-coupling reactions.[80] The palladium catalyst was loaded onto the monolithic support by ion exchange followed by reduction. This supported catalyst promoted Suzuki–Miyaura and Sonogashira couplings, as well as Heck–Mizoroki reactions and transfer hydrogenation reductions of a variety of functional groups (Scheme 4.41). The palladium (0) functionalised monolith was reused in multiple runs for the Heck reaction [Scheme 4.39, eqn (1)] with a decrease in activity from >90% to 71% after seven runs. The Sonogashira coupling [Scheme 4.41, eqn (2)] could be accomplished without the addition of a copper co-catalyst, minimising the risk of blocking the reactor. However, a slight drawback of these reactions in this continuous flow is the long reaction

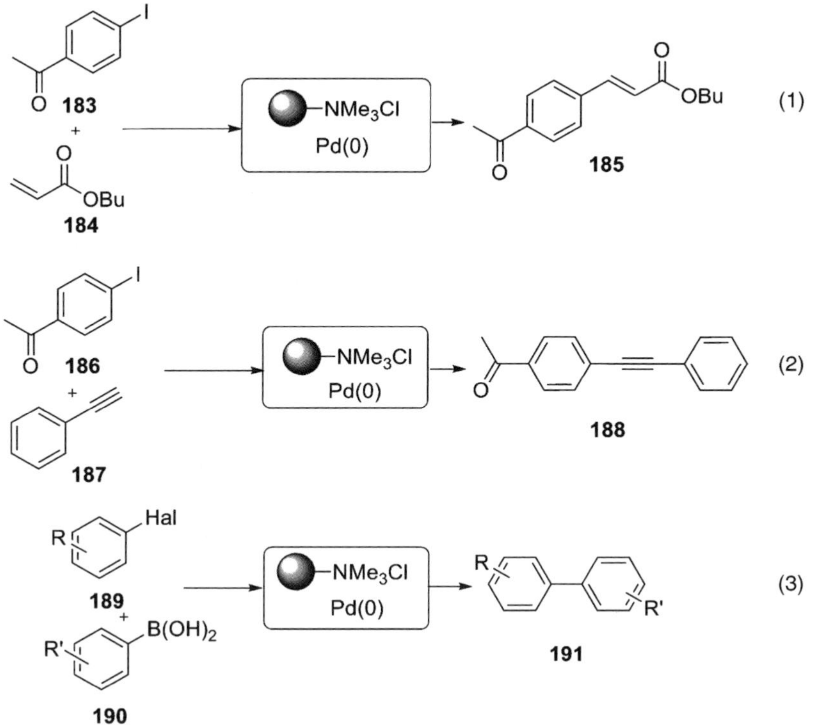

Scheme 4.41 Palladium-monolith catalysed reactions.

times required (2.5–24 h), with recirculation of the reaction solution, and so consequently rather low throughputs were achieved.

A monolithic cartridge derivatised with palladium (0) was also developed by Ley and co-workers for facilitating Heck reactions in EtOH.[81] Using super-heated conditions, the reactions were completed within about 25 min in excellent yields of the Heck products. The monolithic reactor proved to be very robust for consecutive use and was reusable up to 25 times without the need for catalyst regeneration.

The use of porous monolithic materials as a support can be advantageous due to the relatively fast mass transfer between these media with the liquid phase.[82] A monolithic support functionalised with a *N*-heterocyclic carbene/ruthenium metathesis catalyst **193** was used for flow-through ring closing and ring opening metathesis polymerisations reactions (Scheme 4.42). Catalyst bleeding from the monolithic column was limited; metathesis products were obtained with less than 70 parts per million (ppm) of ruthenium.[83]

Poly-urea-encapsulated palladium (II) acetate (PdEnCat)[84] has also been used for Suzuki–Miyaura reactions in continuous flow.[85] Tetra-butylammonium methoxide (Bu$_4$NOMe) was found to be the optimal base for the reaction and quantitative yields were obtained after only four minutes

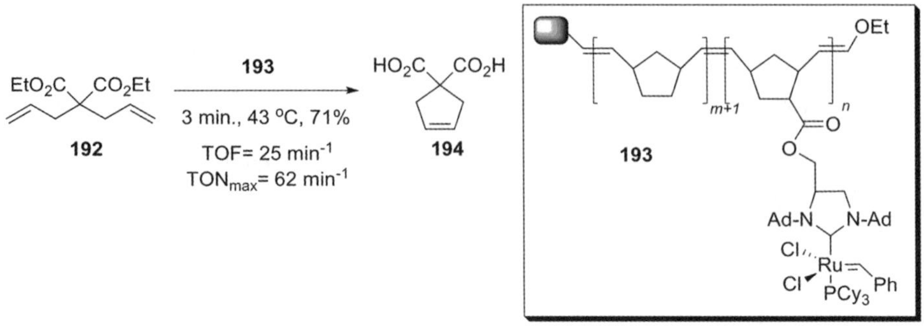

Scheme 4.42 Ruthenium-immobilised monoliths for RCM (Ad = adamantyl).

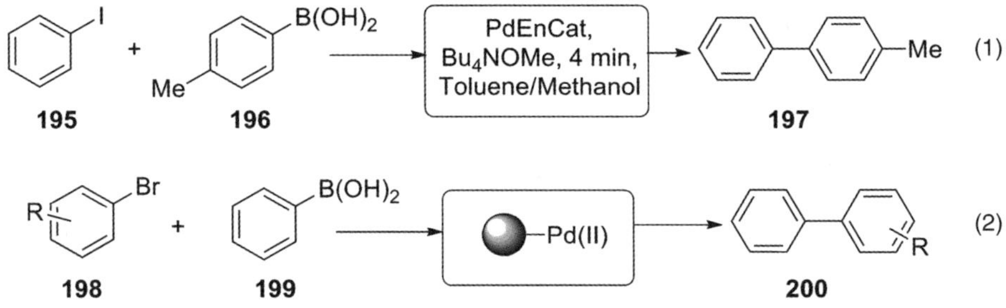

Scheme 4.43 Continuous-flow Suzuki–Miyaura coupling [PdEnCat = polyurea-encapsulated palladium (ii) acetate].

[Scheme 4.43, eqn (1)] . A similar example of Suzuki–Miyaura reactions in flow was reported by Styring and colleagues [Scheme 4.43, eqn (2)].[86] A polymer-supported salen-type palladium complex was used as a catalyst for the coupling of phenylboronic acid and various aromatic halides. Conversions were usually good, but catalyst required some initiation time before displaying full activity.

An ingenious method for the preparation of a membrane-enclosed palladium–phosphine complex inside a microreactor was described by Uozumi and colleagues.[87] A solution of poly(acrylamide)-triarylphosphine in ethyl acetate and an aqueous solution of $(NH_4)_2PdCl_4$ was introduced simultaneously at opposite sides of a microchannel resulting in a parallel, biphasic laminar flow. The formation of the polymeric membrane was immediate, with precipitation at the contacting interface between the two phases. The Suzuki–Miyaura cross-coupling reaction of aryl halides with boronic acids was chosen as a benchmark reaction. Solutions of aryl iodide dissolved in an ethyl acetate/2-propanol solvent mixture and aqueous solutions of aryl boronic acid and sodium carbonate were injected into opposite sides of the membrane-divided reactor. Fine tuning of the reaction conditions included the adjustment of flow rate of each individual flow stream. Under optimal conditions, the reaction was completed

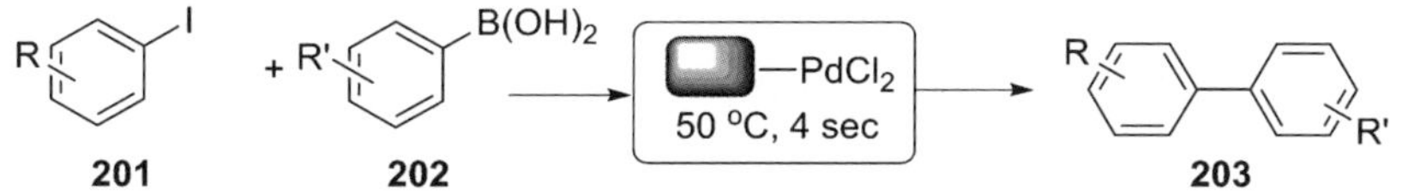

Scheme 4.44 Suzuki–Miyaura couplings using membrane-bound palladium catalyst [= poly(acrylamide)-triarylphosphine].

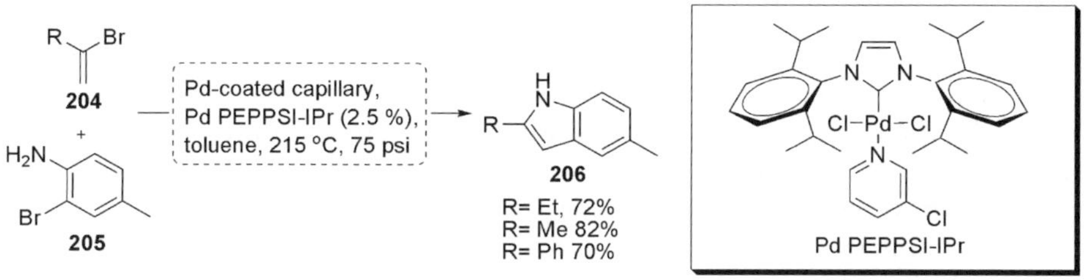

Scheme 4.45 Microwave-assisted tandem aryl amination Heck reactions in flow.

after 4 s residence time at 50 °C with excellent yields of the desired biaryl products (Scheme 4.44).

A tandem amination–intramolecular Heck transformation was developed for the synthesis of indole derivatives using a microflow system consisting of a capillary-size reactor contained within a microwave cavity.[11] Optimal conditions for the preparation of indole derivatives required the simultaneous use of a palladium metal film coated capillary with a homogenous catalyst, Pd PEPPSI-IPr complex (PEPPSI = pyridine-enhanced pre-catalyst preparation stabilisation and initiation); either catalyst systems alone resulted in no conversion. It was speculated that both catalysts may play a role in each of the two steps of the transformation (Scheme 4.45).

4.8 Industrial Uses and Perspectives

Interest in the use of continuous flow microreactor technology in the pharmaceutical and fine chemicals industries continues to grow.[88] With the potential for lower operational costs and enhanced process safety, the application of this technology at an early R&D stage and for intensified processing is, while not yet widespread, slowly advancing. The use of continuous flow microreactors can facilitate the scale-up of established procedures to continuous processing and circumvent time-consuming adjustments of synthetic protocols. The implementation of microflow devices for high throughput screening and synthesis can also have a significant impact on the drug discovery process, as reactions can be performed in very short reaction times and with very small amounts of material. Overall, the 'time to market' can be drastically shortened, rendering the entire process cycle more economical and efficient.

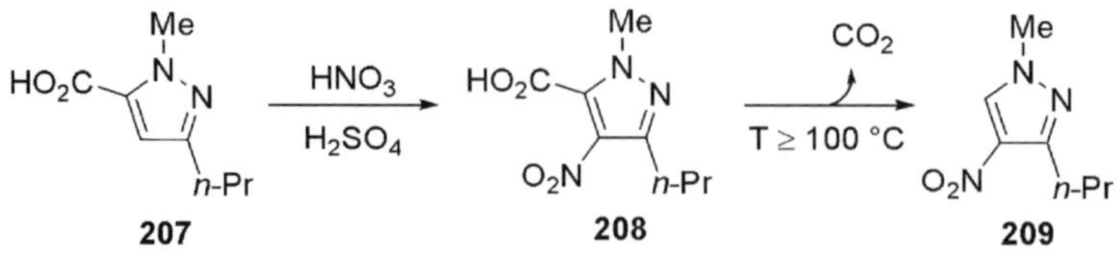

Scheme 4.46 Nitration of substituted pyrazole-5-carboxylic acid **207** in a stainless steel CPC reactor.

Grignard additions,[89a] heterocycle synthesis,[89] halogen–lithium exchange and subsequent additions to nucleophiles,[89b,c] diazotations and coupling reactions,[90] and oxidative dehydrogenations[91] are some examples of the reactions that have been investigated in a process development context. Aromatic nitration in microreactors remains a benchmark example of the advantages to be gained from microflow synthesis (Scheme 4.46).[92]

The protocol for improved processing of electrophilic nitrations of aromatic compounds, which is generally challenging on a large scale when taking safety issues into consideration, was impressively demonstrated in a stainless steel microreactor (CPC CYTOS® lab system, internal volume 70 mL). The nitration of pyrazole-5-carboxylic acid **207**, an important intermediate in the synthesis of Sildenafil®, required extensive temperature control to prevent an undesired exothermic decarboxylation, which was a significant issue in batch remedied only by the slow dosage of nitrating agent (Scheme 4.46). Performing the synthesis in the continuous flow microfluidic device on the other hand permitted simultaneous addition of substrate and reagents, and enabled precise temperature control for the controlled formation of nitropyrazole **208**. The process was operated at a throughput rate of $5.5\,\mathrm{g\,h^{-1}}$ (residence time 35 min), providing the desired product with an isolated yield of 73%.

The prevention of exothermic runaways is a major concern in process development due to safety issues. Rapid heat evolution and the development of 'hotspots' can also have adverse effects on reaction selectivities and can lead to unwanted side-reactions and by-products. For example, exotherms in radical polymerisations can dramatically influence the molecular weight and dispersity of the obtained polymer. Iwasaki and Yoshida developed a stainless steel pilot plant microreactor system using the numbering-up concept (see Section 4.3) to perform radical polymerisations of methyl methacrylate (MMA) **210** and butyl acrylate (BA) **211** using azobisisobutyronitrile (AIBN) as the radical initator (Scheme 4.47).[92] The careful monitoring of reaction temperature, flow rate and flow uniformity allowed for controlled polymerisation at 100 °C and selective termination of the reaction by a drastic temperature change to 25 °C. A flow rate of $55\,\mathrm{mL\,h^{-1}}$ of the neat methyl methacrylate **210** provided a yield of 4 kg per week of poly(methyl methacrylate) (PMMA) with a polydispersity index (PDI) (M_w/M_n) of 1.66.[93]

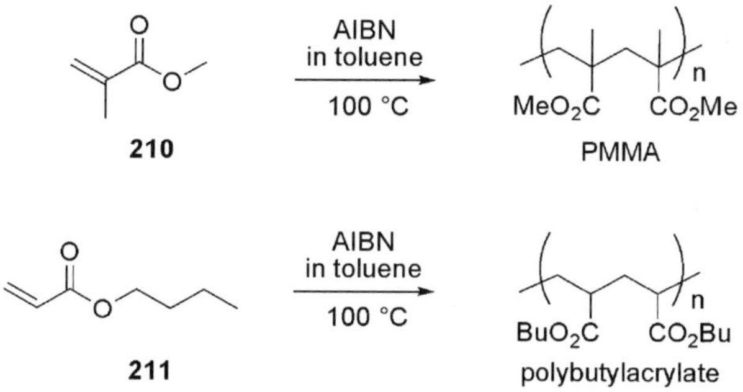

Scheme 4.47 Radical polymerisation of methyl methacrylate **210** and butyl acrylate **211** using AIBN in a stainless steel microreactor.

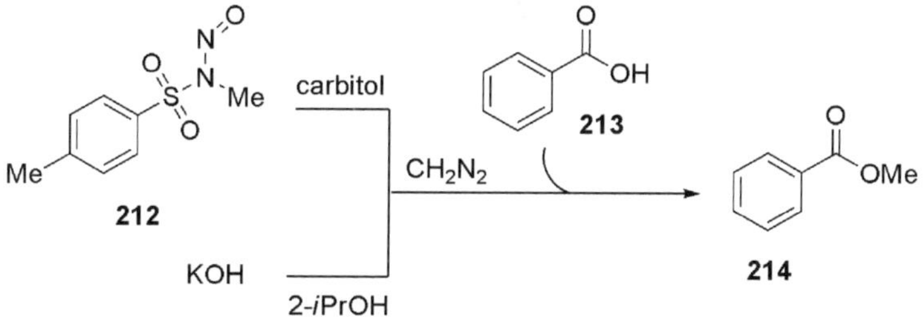

Scheme 4.48 Continuous synthesis of diazomethane from Diazald (**212**).

Diazomethane is a reagent that is frequently avoided even on small laboratory-scale due to its very unattractive characteristics: it is explosive, toxic and carcinogenic, and its low boiling point makes it difficult to handle.[94] Obviously, for the same reasons it cannot usually be considered for industrial use.[95] There are a few reports on continuous diazomethane generation, where the diazomethane is reacted *in situ* to keep overall concentrations low.[96] Proctor and Warr describe a process developed for the synthesis of a key intermediate for the HIV protease inhibitor nelfinavir mesylate.[96a] Diazald (*N*-methyl-*N*-nitroso-*p*-toluene sulphonamide) was treated with potassium hydroxide to generate diazomethane which was purged from the solution with nitrogen as a carrier gas. The reactor was continually drained to ensure steady-state conditions such that the inventory of diazomethane never exceeded 80 g. The process, however, was efficient enough to generate 50–60 tonnes of diazomethane annually. The diazomethane was continuously reacted with substrates inside a packed column. Stark *et al.* also devised a flow method for the generation of diazomethane from Diazald using the transformation of benzoic acid to its methyl ester as a model reaction (Scheme 4.48).[96b]

The Grignard exchange reaction between ethyl magnesium bromide and bromopentafluorobenzene to form pentafluorophenyl magnesium bromide, an

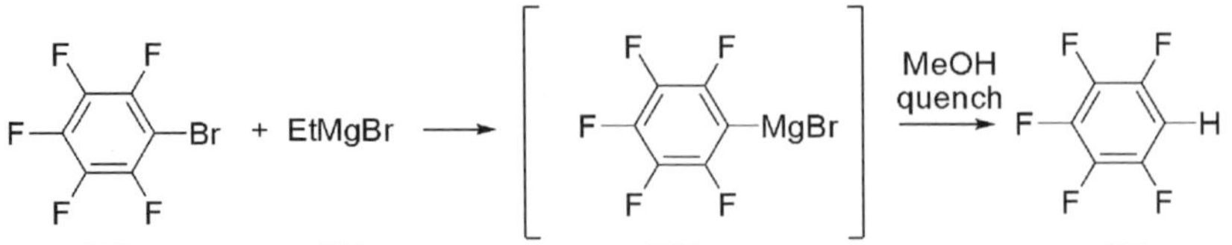

Scheme 4.49   Grignard exchange reaction performed in microflow.

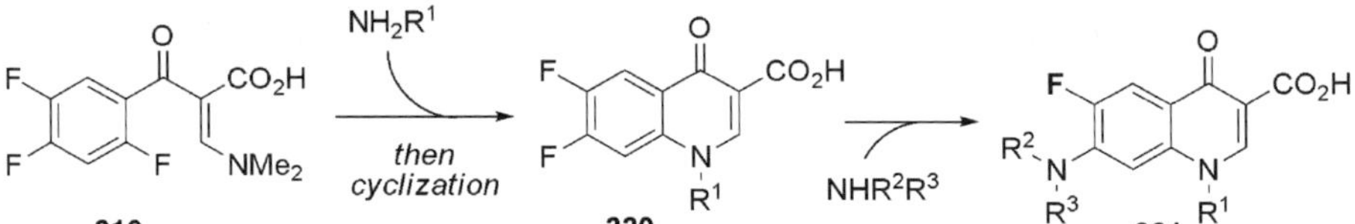

Scheme 4.50   Ciprofloxacin analogue synthesis.

industrially important reagent, was studied in continuous flow.[97] After optimisation of the reaction in smaller and medium-sized systems, a pilot plant scale version consisting of a micromixer connected to a shell and tube micro heat exchanger was developed. Operation for 24 h at 20 °C under optimal flow conditions resulted in the production of 14.7 kg of the pentafluorobenzene (**218**) after quenching with methanol. Temperature, pressure and yields remained consistent throughout the 24 hours of operation. Scheme 4.49

High-throughput screening and combinatorial synthesis can be performed using microreactor devices supported with a high level of automation. Structural diversification is the key to successful lead identification, and so the rapid synthesis of a large collection of compounds is necessary. The use of miniaturised microflow devices is particularly attractive since small quantities of reagents can be used, and reactions can be completed at much faster rates.

For example, a sequential synthesis approach to the fluoroquinoline antibiotic Ciprofloxacin scaffold **221** was developed. Structural diversification was established at two synthetic steps of the synthesis, resulting in the preparation of 21 different Ciprofloxacin analogues (Scheme 4.50).[98] A sequential plug-flow approach was used, where different reagent and solvent pulses were pumped continuously through a single reactor platform. Multiple compounds were generated within a single run without any cross-contamination.

The efficiency and productivity of flow syntheses have been compared to batch syntheses; in some cases, the potential economical impact of flow reactors has been considered.[88c,d] It is certain that many reaction and process limitations can be overcome using microreactor and continuous flow technology, as demonstrated in the examples highlighted here as well as in the previous sections.

References

1. For recent reviews and books on microreactor technology, see: (a) W. Ehrfeld, V. Hessel and H. Löwe, *Microreactors: New Technology for Modern Chemistry,* Wiley-VCH, Weinheim, 2000; (b) P. D. I. Fletcher, S. J. Haswell, E. Pombo-Villar, B. H. Warrington, P. Watts, S. Y. F. Wong and X. L. Zhang, *Tetrahedron,* 2002, **58**, 4735; (c) V. Hessel, S. Hardt and H. Löwe, *Chemical Micro Process Engineering,* Wiley-VCH, Weinheim, 2004; (d) A. M. Thayer, *Chem. Eng. News,* 2005, **83**, 43; (e) K. Geyer, J. D. C. Codée and P. H. Seeberger, *Chem. Eur. J.,* 2006, **12**, 8434; (f) M. Brivio, W. Verboom and D. N. Reinhoudt, *Lab Chip,* 2006, **6**, 329; (g) B. P. Mason, K. E. Price, J. L. Steinbacher, A. R. Bogdan and D. T. McQuade, *Chem. Rev.,* 2007, **107**, 2300; (h) C. Wiles and P. Watts, *Eur. J. Org. Chem.,* 2008, **10**, 1655–1671; (i) T. Wirth, *Microreactors in Organic Synthesis and Catalysis,* Wiley-VCH, Weinheim, 2008.
2. P. Watts and S. J. Haswell, *Chem. Soc. Rev.,* 2005, **34**, 235.
3. K. Jähnisch, V. Hessel, H. Löwe and M. Baerns, *Angew. Chem., Int. Ed.,* 2004, **43**, 406.
4. C. Wiles, P. Watts and S. J. Haswell, *Tetrahedron Lett.,* 2007, **48**, 7362.
5. T. Fukuyama, Y. Hino, N. Katama and I. Ryu, *Chem. Lett.,* 2004, **33**, 1430.
6. B. Ahmed-Omer, J. C. Brandt and T. Wirth, *Org. Biomol. Chem.,* 2007, **5**, 733.
7. T. Fukuyama, T. Rahman, M. Sato and I. Ryu, *Synlett,* 2008, **2**, 151.
8. P. Watts and C. Wiles, *Org. Biomol. Chem.,* 2007, **5**, 727.
9. C. O. Kappe and D. Dallinger, *Nat. Rev. Drug Discovery,* 2006, **5**, 51.
10. J. D. Moseley, P. Lenden, M. Lockwood, K. Ruda, J.-P. Sherlock, A. D. Thomson and J. P. Gilday, *Org. Process Res. Dev.,* 2008, **12**, 30.
11. G. Shore, S. Morin, D. Mallik and M. G. Organ, *Chem J. Eur.,* 2008, **14**, 1351 and references therein.
12. N. Hoffmann, *Chem. Rev.,* 2008, **108**, 1052.
13. E. E. Coyle and M. Oelgemöller, *Photochem. Photobiol. Sci.,* 2008, **7**, 1313.
14. T. Fukuyama, Y. Hino, N. Katama and I. Ryu, *Chem. Lett.,* 2004, **33**, 1430.
15. A. Sugimoto, Y. Sumino, M. Takagi, T. Fukuyama and I. Ryu, *Tetrahedron Lett.,* 2006, **47**, 6197.
16. B. D. A. Hook, W. Dohle, P. R. Hirst, M. Pickworth, M. B. Berry and K. I. Booker-Milburn, *J. Org. Chem.,* 2005, **70**, 7558.
17. M. D. Lainchbury, M. I. Medley, P. M. Taylor, P. Hirst, W. Dohle and K. I. Booker-Milburn, *J. Org. Chem.,* 2008, **73**, 6497.
18. D. R. A. Acke and C. V. Stevens, *Green Chem.,* 2007, **9**, 386.
19. C. Wiles, P. Watts, S. J. Haswell and E. Pombo-Villar, *Org. Process Res. Dev.,* 2004, **8**, 28.
20. E. Garcia-Egidio, V. Spikmans, S. Y. F. Wong and B. II. Warrington, *Lab Chip,* 2003, **3**, 73.

21. O. Flögel, J. D. C. Codee, D. Seebach and P. H. Seeberger, *Angew. Chem., Int. Ed.*, 2006, **45**, 2886.
22. (a) D. M. Ratner, E. R. Murphy, M. D. Jhunjhunwala, A. Snyder, K. F. Jensen and P. H. Seeberger, *Chem. Commun.*, 2005, **5**, 578; (b) K. Fukase, M. Takashina, Y. Hori, D. Tanaka, K. Tanaka and S. Kusumoto, *Synlett.*, 2005, **15**, 2342; (c) K. Geyer and P. H. Seeberger, *Helv. Chim. Acta*, 2007, **90**, 395; (d) F. R. Carrel, K. Geyer, J. D. C. Codée and P. H. Seeberger, *Org. Lett.*, 2007, **9**, 2285.
23. D. Grant, R. Dahl and N. D. P. Cosford, *J. Org. Chem.*, 2008, **73**, 7219.
24. (a) P. Renaud and M. P. Sibi, *Radicals in Organic Synthesis*; Wiley-VCH, Weinheim, Germany, 2001, Vol. 1 and 2; (b) C. Chatgilialoglu, *Organosilanes in Radical Chemistry;* John Wiley & Sons, Chichester, UK, 2004.
25. A. Odedra, K. Geyer, T. Gustafsson, R. Gilmour and P. H. Seeberger, *Chem. Commun.*, 2008, 3025.
26. T. Fukuyama, M. Kobayashi, M. T. Rahman, N. Kamata and I. Ryu, *Org. Lett.*, 2008, **10**, 533.
27. S. C. Roy and S. Adhikari, *Tetrahedron*, 1993, **49**, 8415.
28. G. Maiti, S. Adhikari and S. C. Roy, *Tetrahedron Lett.*, 1994, **35**, 6731.
29. C. Lee, G. Sui, A. Elizarov, C. J. Shu, Y. Shin, A. N. Dooley, J. Huang, A. Daridon, P. Wyatt, D. Stout, H. C. Kolb, O. N. Witte, N. Satyamurthy, J. R. Heath, M. E. Phelps, S. R. Quake and H. Tseng, *Science*, 2005, **310**, 1793.
30. T. L. LaPorte, M. Famedi, J. S. DePue, L. Shen, D. Watson and D. Hsieh, *Org. Process Res. Dev.*, 2008, **12**, 956.
31. T. Schwalbe, V. Autze, M. Hohmann and W. Stirner, *Org. Process. Res. Dev.*, 2004, **8**, 440.
32. (a) H. Usutani, Y. Tomida, A. Nagaki, H. Okamoto, T. Nokami and J. Yoshida, *J. Am. Chem. Soc.*, 2007, **129**, 3046; (b) A. Nagaki, Y. Tomida, H. Usutani, H. Kim, N. Takabayashi, T. Nokami, H. Okamoto and J. Yoshida, *Chem. Asian J.*, 2007, **2**, 1513.
33. A. Nagaki, N. Takabayashi, Y. Tomida and J. Yoshida, *Org. Lett.*, 2008, **10**, 3937.
34. A. K. Ghose, V. N. Viswanadhan and J. J. Wendoloski, *J. Comb. Chem.*, 1999, **1**, 55.
35. A. Basha, M. Lipton and S. M. Weinreb, *Tetrahedron Lett.*, 1977, **18**, 4171–4172.
36. T. Gustafsson, F. Pontén and P. H. Seeberger, *Chem. Commun.*, 2008, 1100.
37. (a) R. Lan, Q. Liu, P. Fan, S. Lin, S. R. Fernando, D. McCallion, R. Pertwee and A. Makriyannis, *J. Med. Chem.*, 1999, **42**, 769; (b) V. K. Kotagiri, S. Suthrapu, J. M. Reddy, C. P. Rao, V. Bollugoddu, A. Bhattacharya and R. Bandichhor, *Org. Process Res. Dev.*, 2007, **11**, 910.
38. L. Kleinberg, S. A. Grossman, K. Carson, G. Lesser, A. O'Neill, J. Pearlman, P. Phillips, T. Herman and M. Gerber, *J. Clin. Oncol.*, 2002, **20**, 3149.

39. (a) K. Omura, A. K. Sharma and D. Swern, *J. Org. Chem.*, 1976, **41**, 957; (b) K. Omura and D. Swern, *Tetrahedron*, 1978, **34**, 1651.

40. (a) T. Kawaguchi, H. Miyata, K. Ataka, K. Mae and J. Yoshida, *Angew. Chem., Int. Ed.*, 2005, **44**, 2413; (b) J. J. M. van der Linden, P. W. Hilberink, C. M. P. Kronenburg and G. J. Kemperman, *Org. Process. Res. Dev.*, 2008, **12**, 911.

41. X. Zhang, S. Stefanick and F. J. Villani, *Org. Process. Res. Dev.*, 2004, **8**, 455.

42. L. Ducry and D. M. Roberge, *Angew. Chem., Int. Ed.*, 2005, **44**, 7972.

43. (a) K. Müller, C. Faeh and F. Diederich, *Science*, 2007, **317**, 1881; (b) K. L. Kirck, *Org. Process Res. Dev.*, 2008, **12**, 305.

44. R. P. Singh and J. M. Shreeve, *Synthesis*, 2002, 2561.

45. D. S. Negi, L. Köppling, K. Lovis, R. Abdallah, J. Geisler and U. Budde, *Org. Process Res. Dev.*, 2008, **12**, 345.

46. T. Gustafsson, R. Gilmour and P. H. Seeberger, *Chem. Commun.*, 2008, 3022.

47. M. Baumann, I. R. Baxendale and S. V. Ley, *Synlett*, 2008, 2111.

48. (a) D. Belder, *Angew. Chem., Int. Ed.*, 2005, **44**, 3521; (b) H. Song, D. L. Chen and R. F. Ismagilov, *Angew. Chem., Int. Ed.*, 2006, **45**, 7336.

49. C. De Bellefon, M. Tanchoux, S. Caravieilhes, P. Grenouillet and V. Hessel, *Angew. Chem., Int. Ed.*, 2000, **39**, 3442.

50. K. Mikami, M. Yamanaka, M. N. Islam, K. Kudo, N. Seino and M. Shinoda, *Tetrahedron*, 2003, **59**, 10593.

51. M. Mikami, M. N. Islam, M. Yamanaka, Y. Itoh, M. Shinoda and K. Kudo, *Tetrahedron Lett.*, 2004, **45**, 3681.

52. M. Ueno, H. Hisamoto, T. Kitamori and S. Kobayashi, *Chem. Commun.*, 2003, 936.

53. H. Okamoto, *Chem. Eng. Technol.*, 2006, **29**, 504.

54. B. Ahmed, D. Barrow and T. Wirth, *Adv. Synth. Catal.*, 2006, **348**, 1043.

55. G. N. Doku, W. Verboom, D. N. Reinhoudt and A. van den Berg, *Tetrahedron*, 2005, **61**, 2733–2742.

56. (a) R. D. Chambers and R. C. H. Spink, *Chem. Commun.*, 1999, 883; (b) K. Jähnisch, M. Baerns, V. Hessel, W. Ehrfeld, V. Haverkamp, H. Löwe, C. Wille and A. Guber, *J. Fluorine Chem.*, 2000, **105**, 117; (c) R. D. Chambers, D. Holling, R. C. H. Spink and G. Sandford, *Lab Chip*, 2001, **1**, 132; (d) R. D. Chambers, M. A. Fox and G. Sandford, *Lab Chip*, 2005, **5**, 1132; (e) N. de Mas, A. Günther, M. A. Schmidt and K. F. Jensen, *Ind. Eng. Chem. Res.*, 2003, **42**, 698; (f) R. D. Chambers, M. A. Fox, D. Holling, T. Nakano, T. Okayoe and G. Sandford, *Chem. Eng. Technol.*, 2005, **28**, 344.

57. (a) D. Wehle, M. Dejmek, J. Rosenthal, H. Ernst, D. Kampmann, S. Trautschold and R. Pechatschek, DE 10036603 A1, 2000; (b) H. Ehrich, D. Linke, K. Morgenschweis, M. Baerns and K. Jähnisch, *Chimia*, 2002, **56**, 647.

58. J. Antes, T. Tuercke, E. Marioth, F. Lechner, M. Scholz, F. Schnürer, H. H. Krause and S. Löbbecke, in *IMRET 5: Proceedings of the Fifth*

International Conference on Microreaction Technology, (ed. M. Matlosz, W. Ehrfeld and J. P. Baselt), Springer, Berlin, 2001, p. 446.

59. (a) K. Jähnisch and M. Baerns, in DD 10257239.9, Germany 2002; (b) R. C. R. Wootton, R. Fortt and A. J. De Mello, *Org. Process. Res. Dev.*, 2002, **6**, 187; (c) Y. Wada, M. A. Schmidt and K. F. Jensen, *Ind. Eng. Chem. Res.*, 2006, **45**, 8036; (d) R. A Bourne, X. Han, A. O. Chapman, N. J. Arrowsmith, H. Kawanami, M. Poliakoff and M. W. George, *Chem. Commun.*, 2008, 4457.

60. (a) P. W. Miller, N. J. Long, A. J. De Mello, R. Villar, J. Passchier and A. Gee, *Chem. Commun.*, 2006, 546; (b) E. R. Murphy, J. R. Martinelli, N. Zaborenko, S. L. Buchwald and K. F. Jensen, *Angew. Chem., Int. Ed.*, 2007, **46**, 1734.

61. T. Fukuyama, R. Yamaura and I. Ryu, *Can. J. Chem.*, 2005, **83**, 711.

62. G. Shi, F. Hong, Q. Liang, H. Fang, S. Nelson and S. G. Weber, *Anal. Chem.*, 2006, **78**, 1972.

63. E. Comer and M. G. Organ, *Chem. Eur. J.*, 2005, **11**, 7223.

64. S. Liu, T. Fukuyama, M. Sato and I. Ryu, *Org. Process Res. Dev.*, 2004, **8**, 477.

65. For further information, see ref. 56b and www.imm-mainz.de/seiten/en.

66. For recent reports on multiphasic hydrogenations see: (a) M. W. Losey, M. A. Schmidt and K. F. Jensen, *Ind. Eng. Chem. Res.*, 2001, **40**, 2555; (b) M. W. Loosey, R. J. Jackman, S. L. Firebaugh, M. A. Schmidt and K. F. Jensen, *J. Microelectromech. Syst.*, 2002, **11**, 709; (c) K. K. Yeong, A. Gavriilidis, R. Zapf and V. Hessel, *Catal. Today*, 2003, **81**, 641; (d) J. Kobayashi, Y. Mori, K. Okamoto, R. Akiyama, M. Ueno, T. Kitamori and S. Kobayashi, *Science*, 2004, **304**, 1305; (e) J. Kobayashi, Y. Mori and S. Kobayashi, *Chem. Commun.*, 2005, 2567; (f) S. Saaby, K. R. Knudsen, M. Ladlow and S. V. Ley, *Chem. Commun.*, 2005, 2909; (g) N. Yoswathananont, K. Nitta, Y. Nishiuchi and M. Sato, *Chem. Commun.*, 2005, 40; (h) R. Jones, L. Gödörházy, D. Szalay, L. Ürge and F. Darvas, *QSAR Comb. Sci.*, 2005, **4**, 722; (i) B. Desai and O. C. Kappe, *J. Comb. Chem.*, 2005, **7**, 641; (j) G. Szöllösi, B. Hermán, F. Fülép and M. Bartók, *React. Kinet. Catal. Lett.*, 2006, **88**, 391; (k) R. Jones, L. Gödörházy, N. Vargs, D. Szalay, L. Ürge and F. Darvas, *J. Comb. Chem.*, 2006, **8**, 110.

67. For recent reports on multiphasic carbonylations in microdevices, see: (a) P. W. Miller, N. J. Long, A. J. De Mello, R. Vilar, H. Audrain, D. Bender, J. Passchier and A. Gee, *Angew. Chem., Int. Ed.*, 2007, **46**, 2875; (b) C. Csajági, B. Borcsek, K. Niesz, I. Kovács, Z. Székelyhidi, Z. Bajkó, L. Ürge and F. Darvas, *Org. Lett.*, 2008, **10**, 1589.

68. For reports on triphasic oxidations see: (a) F. Loeker and W. Leitner, *Chem. Eur. J.*, 2000, **6**, 2011; (b) G. Jenzer, T. Mallat and A. Baiker, *Catal. Lett.*, 2001, **73**, 5; (c) G. Jenzer, M. S. Schneider, R. Wandeler, T. Mallat and A. Baiker, *J Catal*, 2001, **197**, 141; (d) G. Jenzer, T. Mallat, M. Maciejewski, F. Eigenmann and A. Baiker, *Appl. Catal., A*, 2001, **208**, 125; (e) Z. Hou, N. Theyssen, A. Brinkmann and W. Leitner, *Angew.*

     Chem., Int. Ed., 2005, **11**, 1346;   (f) N. Theyssen, Z. Hou and W. Leitner, *Chem. Eur. J.*, 2006, **12**, 3401.

69. J. H. Clark and S. J. Tavener, *Org. Process Res. Dev.*, 2007, **11**, 149.

70. (a) Y. Orito, S. Imai and S. Niwa, *J. Chem. Soc. Jpn.*, 1979, **8**, 1118; (b) Y. Orito, S. Imai and S. Niwa, *J. Synth. Org. Chem.*, 1979, **37**, 173.

71. (a) P. A. Meheux, A. Ibbotson and P. B. Wells, *J. Catal.*, 1991, **128**, 387–396; (b) N. Künzle, R. Hess, T. Mallat and A. Baiker, *J. Catal.*, 1999, **186**, 239.

72. For further information, see http://thalesnano.com.

73. P. H. Toy and M. Shi, *Tetrahedron*, 2005, **61**, 12025.

74. I. R. Baxendale, C. M. Griffiths-Jones, S. V. Ley and G. K. Tranmer, *Synlett*, 2006, **3**, 427.

75. I. R. Baxendale, S. V. Ley, C. D. Smith and G. K. Tranmer, *Chem. Commun.*, 2006, 4835.

76. M. Baumann, I. R. Baxendale, S. V. Ley, C. D. Smith and G. K. Tranmer, *Org. Lett.*, 2006, **8**, 5231.

77. (a) I. R. Baxendale, J. Deeley, C. M. Griffith-Jones, S. V. Ley, S. Saaby and G. K. Tranmer, *Chem. Commun.*, 2006, 2566; (b) S. V. Ley, O. Schucht, A. W. Thomas and P. J. Murray, *J. Chem. Soc., Perkin Trans. 1*, 1999, **10**, 1251.

78. C. D. Smith, I. R. Baxendale, S. Lanners, J. J. Hayward, S. C. Smith and S. V. Ley, *Org. Biomol. Chem.*, 2007, **5**, 1559.

79. (a) C. D. Smith, I. R. Baxendale, G. K. Tranmer, M. Baumann, S. C. Smith, R. A. Lewthwaite and S. V. Ley, *Org. Biomol. Chem.*, 2007, **5**, 1562; (b) M. Baumann, I. R. Baxendale, S. V. Ley, N. Nikbin, C. S. Smith and J. P. Tierney, *Org. Biomol. Chem.*, 2008, **6**, 1577; (c) M. Baumann, I. R. Baxendale, S. V. Ley, N. Nikbin and C. S. Smith, *Org. Biomol. Chem.*, 2008, **6**, 1587.

80. (a) W. Solodenko, H. Wen, S. Leue, F. Stuhlman, G. Sourkouni-Argirusi, G. Jas, H. Schönfeld, U. Kunz and A. Kirschning, *Eur. J. Org. Chem.*, 2004, **1**, 3601; (b) K. Mennecke, W. Solodenko and A. Kirschning, *Synthesis*, 2008, **10**, 1589.

81. N. Nikbin, M. Ladlow and S. V. Ley, *Org. Process Res. Dev.*, 2007, **11**, 458.

82. U. Kunz, A. Kirschning, H.-L. Wen, W. Solodenko, R. Cecilia, C. O. Kappe and T. Turek, *Catal. Today*, 2005, **105**, 318.

83. M. Mayr, B. Mayr and M. R. Buchmeiser, *Angew. Chem., Int. Ed.*, 2001, **40**, 3839.

84. C. Ramarao, S. V. Ley, S. C. Smith, I. M. Shirley and N. DeAlmeida, *Chem. Commun.*, 2002, 1132.

85. C. K. Y. Lee, A. B. Holmes, S. V. Ley, I. F. McConvey, B. Al-Duri, G. A. Leeke, R. C. D. Santos and J. P. K. Seville, *Chem. Commun.*, 2005, 2175.

86. N. T. S. Phan, J. Khan and P. Styring, *Tetrahedron*, 2005, **61**, 12065.

87. Y. Uozumi, Y. M. A. Yamada, T. Beppu, N. Fukuyama, M. Ueno and T. Kitamori, *J. Am. Chem. Soc.*, 2006, **128**, 15994.

88. (a) S. Tghavi-Moghadam, A. Kleemann and K. G. Golbig, *Org. Process Res. Dev.*, 2001, **8**, 652; (b) See ref. 1(i); (c) See ref. 31; (d) D. M. Roberge,

L. Ducry, N. Bieler, P. Cretton and B. Zimmermann, *Chem. Eng. Technol,* 2005, **28**, 318; (e) D. M. Roberge, B. Zimmerman, F. Rainone, M. Gottsponer, M. Eyholzer and N. Kockmann, *Org. Process Res. Dev.*, 2008, **12**, 905.

89. (a) See ref. 41; (b) J. Choe, J. H. Seo, Y. Kwon and K. H. Song, *Chem. Eng. J.*, 2008, **135**, S17; (c) N. Kockmann, M. Gottsponer, B. Zimmermann and D. M. Roberge, *Chem. Eur. J.*, 2008, **14**, 7470.

90. C. Wille, H. P. Gabski, T. Haller, H. Kim, L. Unverdorben and R. Winter, *Chem. Eng. J.*, 2004, **101**, 179.

91. O. Wörz, K. P. Jäckel, T. Richter and A. Wolf, *Chem. Eng. Sci.*, 2001, **56**, 1029.

92. (a) G. Panke, T. Schwalbe, W. Stirner, S. Taghavi-Moghadam and G. Wille, *Synthesis*, 2003, **18**, 2827; (b) See http://www.cpc-net.com and http://www.acclavis.com/Layout%202.htm for further information.

93. T. Iwasaki and J. Yoshida, *Macromolecules*, 2005, **38**, 1159.

94. C. D. Gutsche, *Org. Reactions*, 1954, **8**, 391.

95. Aerojet-General Corporation has reported two procedures for the large-scale generation of diazomethane (50 g to 25 kg as ethereal solutions), see: (a) T. G. Archibald, D.-S. Huang, M. H. Pratton and J. C. Bernard, US Pat., 5 817 778, 1998; (b) T. G. Archibald, J. C. Bernard and R. F. Harlan, US Pat., 5 854 405, 1998.

96. (a) L. D. Proctor and A. J. Warr, *Org. Process Res. Dev.*, 2002, **6**, 884; (b) M. Struempel, B. Ondruschka, R. Daute and A. Stark, *Green Chem.*, 2008, **10**, 41.

97. H. Wakami and J. Yoshida, *Org. Process Res. Dev.*, 2005, **9**, 787.

98. T. Schwalbe, D. Kadzimirsz and G. Jas, *QSAR Comb. Sci.*, 2005, **24**, 758.

Flow Processes in Non-Conventional Media

TÂNIA QUINTAS AND DAVID J. COLE-HAMILTON

EaStCHEM, School of Chemistry, University of St Andrews, Fife, KY16 9ST, UK

5.1 The Need for Alternative Solvents in Flow Catalysis

5.1.1 Homogeneous *vs.* Heterogeneous Catalysis

A chemical process involving reactants and a catalyst is denominated catalysis. If during the catalytic process both the catalyst and the reactants are in the same phase, mostly the liquid phase, it is called homogeneous catalysis. Otherwise, it is called heterogeneous catalysis.

Although homogeneous catalysis has very clear advantages when compared to heterogeneous catalysis, many of the known homogeneous catalytic systems are still not being commercialised because of problems concerning the separation of the catalyst from the product, the retention of the catalyst in the reaction medium and the use of organic solvents.[1,2]

The advantages of homogeneous catalysts over heterogeneous catalysts include:

- their higher specificity;
- the availability of all the catalytic centres because of the dissolution of the catalyst in a solvent;

RSC Green Chemistry No. 5
Chemical Reactions and Processes under Flow Conditions
Edited by S.V. Luis and E. Garcia-Verdugo
© The Royal Society of Chemistry 2010
Published by the Royal Society of Chemistry, www.rsc.org

- reproducibility;
- controllability.

Despite this, heterogeneous catalysts can be a cheaper option for the chemical industry because of their higher thermal stability, the lack of need for a large quantity of suitable solvents but, most of all, the easy separation of the very expensive catalyst from the product and its easy reuse. Table 5.1 provides a simple way of illustrating the advantages and disadvantages of homogeneous versus heterogeneous catalysis.

Society in general, and scientists in particular, are starting to be aware of the dangers and consequences caused by the damage that is being done to our environment and efforts are being made in order to create and/or preserve a better world and environment for all of us. Alongside the catalyst separation problem, there are also the many environmental problems that are starting to concern and affect society. It is well-known that volatile organic compound (VOC) vapours can be released by organic solvents to the atmosphere and that their use on a large scale by the chemical industry is not benign to the environment.

Over the last few decades, the chemical industry has been trying to create, develop and substitute better and cleaner industrial processes for existing ones. An important part of this research involves the substitution of common solvents by alternatives, which not only reduce emissions of VOCs to the atmosphere, but also offer processing advantages in terms of allowing reactions that would otherwise be carried out in batch mode to be made continuous.

Homogeneous catalysis is usually carried out in batch or batch continuous reactions. Batch reactions have the disadvantage that the reactor must be opened at the end of the reaction; the product, catalyst and solvent must be separated and, if possible, the catalyst and solvent recycled. This leads to very significant downtime and low production efficiency.

Batch continuous reactions operate continuously (Figure 5.1) with substrates being fed into a reactor—usually a continuously stirred tank reactor (CSTR). Part of the reaction solution flows into a separator where the product, solvent and catalyst are separated in a continuous process—usually distillation or phase separation. The phase or residue containing the catalyst is then fed back into the reactor whilst the products are removed for further processing. The

Table 5.1   Homogeneous *vs.* heterogeneous catalysis (reproduced with permission from ref. 3 © Springer).

	Homogenous	*Heterogeneous*
Activity	+ + +	−
Selectivity	+ + +	+
Catalyst description	+ +	−
Catalyst recycling	−	+ + +
TON (turnover number)	+	+ + +
Quantity of catalyst	+ +	+ + +

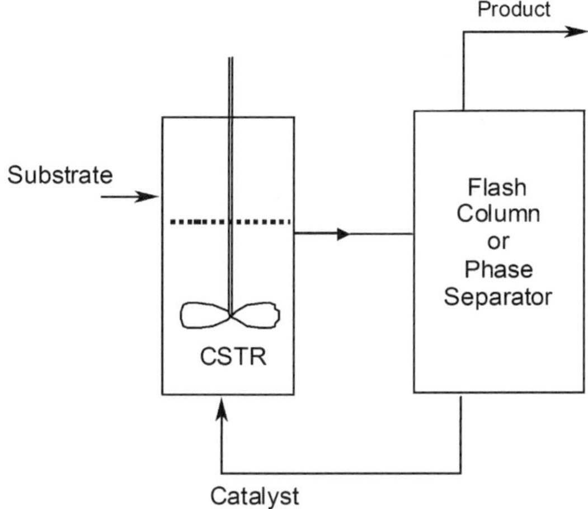

Figure 5.1 Schematic diagram of a batch continuous reactor.

main disadvantage of such batch continuous processes is that some of the catalyst is in the separator or recycling loop. Not only does this mean that a larger amount of catalyst than is necessary to sustain the reaction at the desired rate must be used, but also that it is under conditions very different from those in the catalytic reactor for which it has been optimised. Often the conditions in the separator can be harsh (high temperature, low pressure) and can lead in some cases to decomposition or to precipitation. In the worst cases, these problems may lead to clogging of the recycling loop and very considerable downtime.

5.2 Continuous Flow Processing using Homogeneous Catalysis

To address these disadvantages, it would be ideal if homogeneously catalysed reactions could be carried out under conditions where the substrate flows through the catalyst solution and the product flows out of the reactor, with the catalyst remaining in the reactor at all times (Figure 5.2). Such a system would be very similar to those used for heterogeneous catalysts.

Such flow systems are possible if all the reagents are volatile and if all the products can be distilled from the reactor containing a non-volatile solvent and a non-volatile catalyst. An example is the commercial hydroformylation of propene to butanals. This is carried out using Rh/PPh_3 catalysts with aldol condensation products of the formed aldehydes acting as the solvent. Propene, CO and H_2 are fed into the reactor in the gas phase whilst the product is also removed in a continuously flowing process in the gas phase. This is possible because the boiling point of butanal is $75\,°C$ and the reaction is carried out at about $100\,°C$.[4] To continue with the example of hydroformylation, this

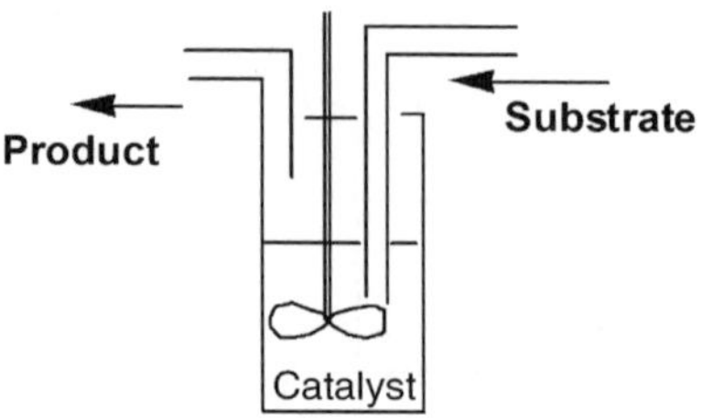

Figure 5.2 Desired process for continuous flow homogeneous catalysis.

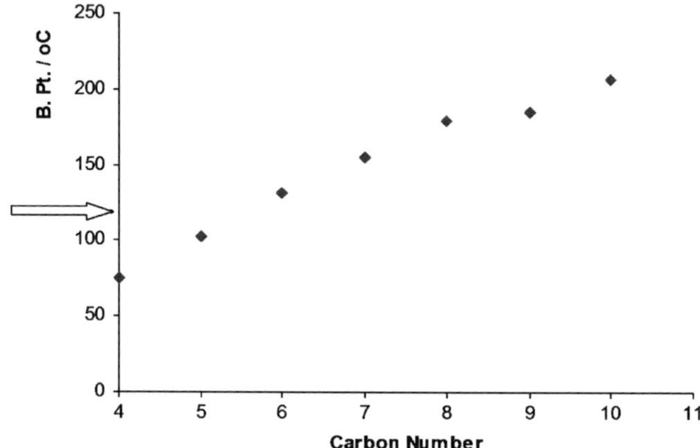

Figure 5.3 Effect of chain length on the boiling point of linear aldehydes.

continuous flow approach is difficult for higher boiling aldehydes derived from longer chain alkenes because the rhodium-based catalyst decomposes at *ca.* 110 °C and most aldehydes boil at higher temperatures than this (Figure 5.3).

Longer chain aldehydes are still manufactured by homogeneous hydro-formylation reactions (though not in continuous flow processes) because they are valuable materials used as plasticisers and in the manufacture of soaps and detergents, and because alkenes are relatively cheap. In most cases, cobalt catalysts are employed sometimes with added alkylphosphines.[4,5] These require more forcing conditions than rhodium and are less selective, either giving much higher amounts of undesired branched products (no phosphine added) or giving significant amounts of alkane side product from hydrogenation of the alkene (phosphine modified; alcohols rather than aldehydes are the predominant products). These reactions are run in batch continuous mode with the product separation involving washes with sodium hydroxide and neutralisation, leading to salt formation. Only very recently has Rh/PPh$_3$ been commercialised by Sasol using technology developed by Kvaerner (now Davey) Process Technology. This is again run in batch continuous mode with the separation being carried out by reduced pressure distillation.[6]

This brief discussion of alkene hydroformylation serves to illustrate some of the difficulties of homogeneous catalysis and the barriers that exist to operation under continuous flow mode. However, recent developments have introduced new types of solvents that do allow homogeneous catalytic reactions to be carried out under genuinely continuous flowing conditions.[2,7] In this chapter we review progress that has been made in the use of non-traditional solvents to allow homogeneous catalysis to be carried out under flow conditions. In addition, we highlight systems where non-conventional solvents have been used in batch continuous processes.

5.3 The Use of Solvents

A solvent is typically described as any material which is capable of dissolving another substance.[8] For many years, industry has been using common organic solvents such as, for example, chloroform, tetrahydrofuran (THF), dichloromethane (DCM) or acetone, with serious consequences to the environment. These consequences come mainly from the fact that solvents are normally used in very large quantities during the chemical processes and also because they are usually volatile liquids (VOCs) that are very hard to contain without spreading to the atmosphere.

Besides being used as reaction media in most chemical processes, solvents are also used as cleaning solutions, to perform a separation or an extraction, or even to take part in product synthesis. Whatever the main aim of the use of a solvent, it is important to choose it properly taking in account its various properties including:

- viscosity;
- polarity;
- solubilising power;
- volatility.

These properties are very important and should be studied as a whole so that, during any chemical process, the solvent involved should be able to dissolve the reagents and reactants completely, be inert to all the reaction conditions, have an appropriate boiling point and, most of all, be easily removed from the other products at the end of the process. Because the easiest ways to remove the solvent at the end of the process are normally by evaporation or distillation, volatile solvents have become popular over the years.

Together with the fact that common solvents emit vapours to the atmosphere, contributing to global warming and sometimes being involved in the depletion of the ozone layer, there are other hazards to take in account when using common solvents, such as their toxicity and flammability. Therefore, it should come as no surprise that these are among the most regulated chemicals.[9,10]

5.3.1 Traditional Solvents *vs.* Non-conventional Solvents

Given the points explained above, researchers and industry have been joining forces together over the last decades to develop new classes of solvents where the main aim is to reduce significantly all the hazards associated with traditional solvents. This can be achieved through the design of new substances such as ionic liquids for which, because of their very diverse nature, the performance can be tuned to be as good as or even better than conventional solvents. Another alternative is to use no solvents at all, the so-called solventless systems.[11] Other successful examples of applications of new reaction media include the use of water (not previously widely applied in the chemical industry), supercritical fluids and fluorous solvents amongst others. Industry and researchers have been working hard to exploit the advantages of these solvents and to implement them in well-known chemical processes. We now review recent progress in the use of ionic liquids and supercritical fluids, highlighting their uses in homogeneous catalysis and especially, where possible, in flow systems.

5.4 Ionic Liquids

5.4.1 The concept of Ionic Liquids

Ionic liquids or molten salts are liquids at room temperature consisting only of ions. The main difference between ionic liquids and common molten salts (*e.g.* NaCl) is the fact that they are liquids at low temperatures.[12] Although the first room temperature ionic liquid, $[EtNH_3][NO_3]$, was discovered in 1914, they were only 'rediscovered' a few decades ago; since then, their properties and potential applications have been increasing enormously[13]. Nowadays ionic liquids are considered as a whole 'new class of solvents' and have been extensively studied and tested in various chemical processes.[12,14–19]

One of their many advantages of ionic liquids compared with traditional solvents is the possibility of combining several anions and cations in order to synthesise several potential ionic liquids. Basically, it is possible to 'design' an ionic liquid to represent the ideal solvent properties for a certain reaction. Amongst important types of cations used in the synthesis of ionic liquids there are, for instance, the dialkyl imidazolium ions, the tetraalkyl ammonium ions, the alkylpyridinium ions or the tetraalkyl phosphonium ions. As for anions, the most common ones are $[AlCl_4]^-$, $[PF_6]^-$, $[BF_4]^-$ and $[Tf_2N]^-$ $(Tf = CF_3SO_2)$ (Figure 5.4). Just to give an idea about the potential scope, it is possible to have around 10^{18} potential ionic liquids only in the imidazolium and pyridinium systems, just by varying the combination of anions and cations.[20]

By changing the nature of the cation or the anion of the ionic liquid, it is possible to change properties such as viscosity and solubility.[19] For instance, by increasing the size of the alkyl chain of the cation composing part of the ionic liquid, it is possible to increase its solvating properties for less polar solutes.[16] This ability to adjust the properties of the ionic liquids through the immense

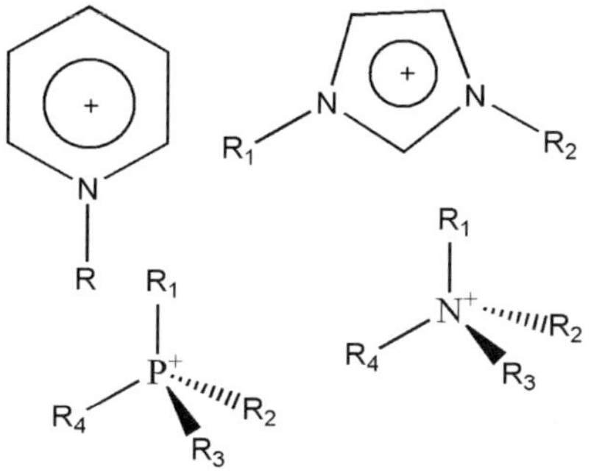

BF_4^-, PF_6^-, $CF_3SO_2^-$, $(CF_3SO_2)_2N^-$,
$CH_3SO_3^-$, $AlCl_4^-$, $Al_2Cl_7^-$, NO_3^-

Figure 5.4 Popular cations (on the left) and anions (on the right) used in the synthesis of ionic liquids.

possibilities of cation–anion combination gave them the well-deserved name of 'designer solvents'.

Two of the greatest advantages of ionic liquids when compared with traditional solvents are their non-measurable vapour pressure and their non-flammability. This potentially makes them environmentally benign as they do not release vapours into the atmosphere. Other factors must also be considered, however, such as environmental persistence, toxicity, corrosivity and the environmental impact of their synthesis. Since there are so many different possible ionic liquids, it is impossible to generalise, but it should be possible to find ionic liquids which will have minimum negative environmental impact throughout their life cycle. The absence of a measurable vapour pressure is very advantageous in terms of product separation from the reaction medium and the catalyst by distillation.[12] Most of the ionic liquids used are also very thermally stable, which allows chemical reactions or separations to be carried out at very high temperatures without negative effects on the ionic liquid.

In summary, some of the properties of ionic liquids which allow them to be considered as true and reliable alternatives to replace traditional solvents and stand as a 'new class of solvents' are:

- tuneable solvation properties;
- non-measurable vapour pressure;
- non-flammable below decomposition temperature;
- high thermal stability;
- very good solvents for a variety of organic and inorganic compounds;
- high electrical conductivity;
- very cheap and easy to prepare (in some cases).

In 1972, Parshall reported the first successful platinum catalysed hydroformylation reaction carried out in an ionic liquid reaction medium.[21] He introduced the use of tetraalkylammonium salts of $GeCl_3^-$ and $SnCl_3^-$ as solvents—then referred to as non-conventional molten salts—in several catalytic reactions because of their non-volatility, thermal stability and the possibility of retaining the metal catalyst and increasing the reaction selectivity.

5.4.2 Continuous Flow Catalysis using Ionic Liquids

5.4.2.1 Liquid–Liquid Biphasic Catalysis

One of the main applications of ionic liquids (ILs) as solvents is in liquid–liquid biphasic catalysis (Figure 5.5). In such systems there are two immiscible solvents but only one of the phases contains the catalyst; this allows very easy separation of the product by decantation and recycling of the catalyst phase.[12,19] The characteristics of these biphasic systems may allow homogeneous systems to behave almost like heterogeneous systems.

Due to their tunable solvating properties, it is possible to 'design' the right ionic liquid to guarantee that the catalyst will be totally dissolved in the ionic liquid phase. The main result of this is the verified reduction in the leaching from the reaction medium into the product phase. In terms of industry, this is a big advantage when considering the price of the metal catalysts, which can now be used for longer periods of time.

The first ionic liquid based process to be used in industry, which was introduced by Eastman Chemical Company in 1998,[22] involved the use of long-chain tetraalkylphosphonium salts in a Lewis acid catalysed isomerisation of monoepoxidised butadiene to dihydrofuran, an important intermediate from which higher added products can be synthesised. In this process, certain very polar low molecular weight oligomeric side products develop. The reaction is carried out in the presence of the phosphonium salt, which has very low polarity and is soluble in heptane. After the volatile products have been distilled from the reactor, a concentrated solution containing the catalyst and the

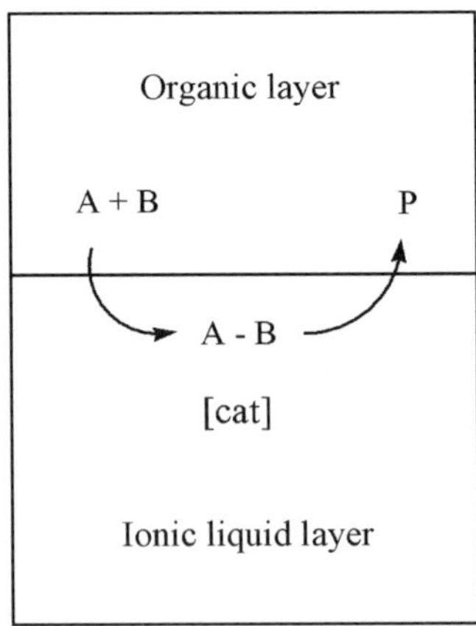

Figure 5.5 Liquid–liquid biphasic catalysis using ionic liquids for catalyst retention.

phosphonium salt is fed continuously to a counter-current extractor where the catalyst, phosphonium salt and polar side products are separated using heptane. The catalyst and phosphonium salt dissolve, but the side products do not. The separated side products are sent for incineration, whilst the non-polar extract is distilled to remove heptane and to leave the catalyst and the phosphonium salt, which are recycled to the reactor. This batch continuous process, which unusually exploits the very low polarity of the liquid phosphonium salt, ran for eight years with minimal losses of phosphonium salt[23] (Figure 5.6).

In 2002, BASF™ introduced another commercial use of ionic liquids, BASIL™ (**B**iphasic **A**cid **S**cavenging utilizing **I**onic **L**iquids) process.[24] Alkoxyphenylphosphines are used as raw materials for the production of photoinitiators to cure

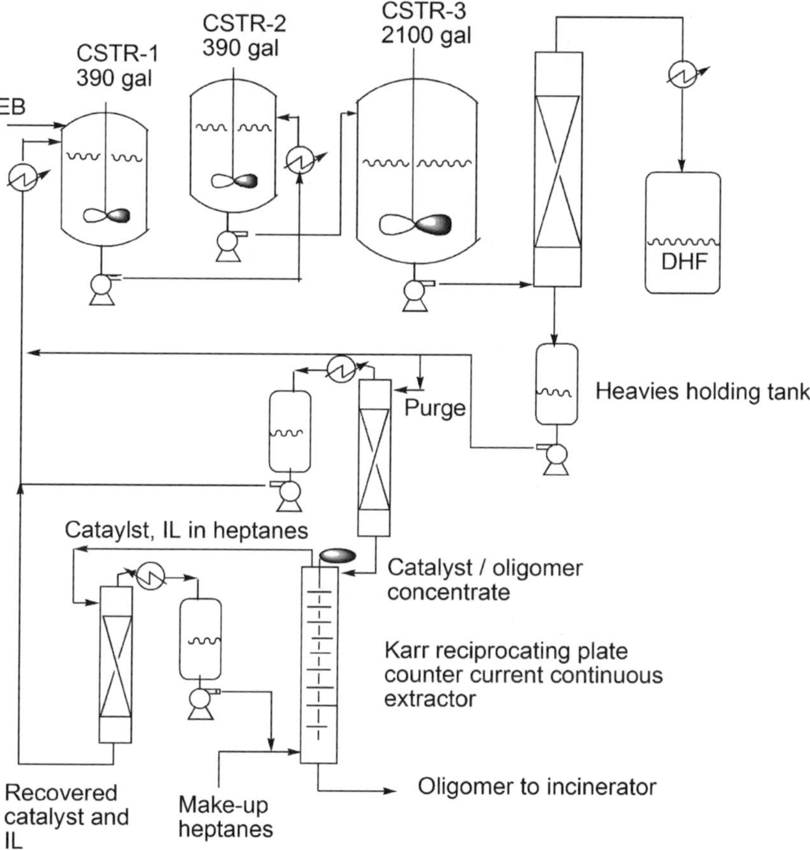

Figure 5.6 Schematic diagram of the reactor and separator used by the Eastman Chemical Company for the isomerisation of epoxybutadiene to dihydrofuran. The catalyst consists of a Lewis acid and an apolar quaternary phosphonium salt.[22] After ref. 23 Reproduced by courtesy of The Catalyst Group.

$$PhPCl_2 \ + \ 2\,EtOH \ + 2\,R_3N \ \longrightarrow \ PhP(OEt)_2 \ + \ 2\,R_3N \ + \ HCl$$

Scheme 5.1   Synthesis of diethylphenylphosphonite.

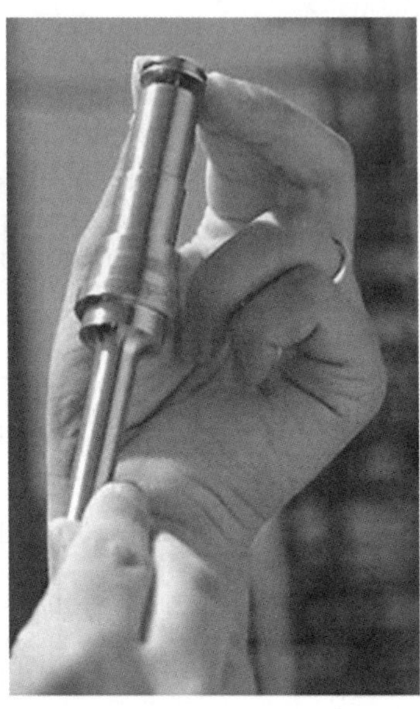

Figure 5.7   The BASIL process is run in a small jet reactor. The plant has a total capacity of $690\,000\,kg\,m^{-3}\,h^{-1}$. Reproduced with permission from ref. 25. © BASF.

coatings and printing inks by exposure to ultraviolet (UV) light. Their synthesis (Scheme 5.1) has the major drawback of also producing HCl as a co-product. Conventionally, tertiary amines can be used to scavenge the HCl, but the thick, non-stirrable slurry resulting from the scavenging reduces the efficiency of the process enormously. It was also necessary to carry out the process in batch mode in order to filter the product after the reactor. The throughput was very low and the process was difficult to operate.

By introducing 1-methylimidazole, an ionic liquid precursor, as the HCl scavenger in the production of the diethyphenylphosphonite, Maase and co-workers[25] greatly improved the rate and efficiency of the process because 1-methyl-imidazolium chloride, which forms, is a colourless mobile liquid that separates from the product. The imidazole has a nucleophilic catalytic action in the process, so the whole reaction can take place in less than one second, leading to an enormous increase in the reaction rate and process productivity (by a factor of 8×10^4 to $690\,000\,kg\,m^{-3}\,h^{-1}$). In 2004, the BASIL process was initiated using a large vessel with the actual reaction being carried out in very small jet reactors (Figure 5.7).[25] It is also possible to regenerate the imidazole and release HCl on heating.

Another good example of these systems is the Dimersol X process, which was developed by the Institut Français du Pétrole (IFP).[26–28] The dimerisation of butene is catalyzed by Ni(II) salt complexes activated with an alkylaluminium co-catalyst. Although it has all the advantages of a homogeneous process, it also comes with the main associated drawback—the big loss of catalyst when carried out continuously. Chauvin, Olivier-Bourbigou and other researchers from the IFP identified a suitable biphasic system by using 1,3-dialkylimidazolium chloroaluminate as a solvent for this process. This ionic liquid creates a good liquid–liquid biphasic system since it is a good solvent for butenes but not for the dimerisation products. The expulsion of the product leads to fewer secondary products. Furthermore, the ionic liquid can also dissolve and stabilise the Ni catalyst without even needing the help of a ligand.

Tests were run using a continuous flow through a CSTR followed by a phase separator (Figure 5.8). For a period of 5500 hours, Raffinate II (industrial feed composed of 70% by weight butenes of which 27% by weight was 1-butene) was fed continuously into the reactor where the ionic liquid layer was mixed vigorously with the organic layer. The mixture then passed to the phase separator, where gravity separation happened very fast and cleanly. The upper organic phase containing the products escaped through an overflow for collection and processing, whilst the lower ionic liquid phase containing the catalyst was recycled back to reactor. After deliberately stopping the tests, the researchers concluded that it was not necessary to add more ionic liquid during the tests and that the Ni leaching had decreased dramatically compared with the homogeneous Dimersol process.

These examples all involve batch continuous processing where part of the catalytic solution is continuously removed for product separation and catalyst recovery (recycling). One possible way in which such catalysts might be used for genuine continuous operations might be to use a counter-current in a loop type reactor. The catalyst dissolved in the ionic liquid would be fed from the bottom

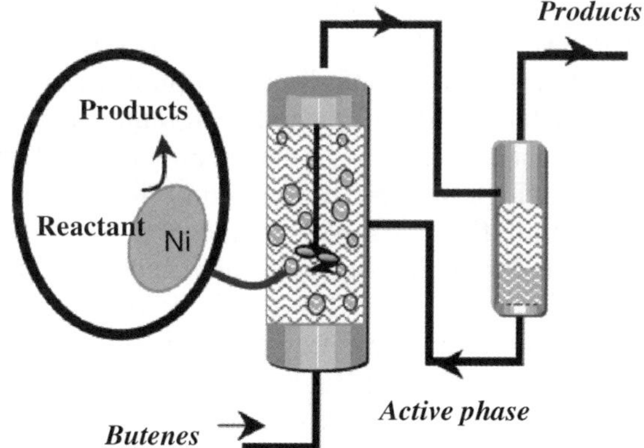

Figure 5.8 Continuous dimerisation of butenes using 1,3-dialkylimidazolium chloroaluminate to immobilise the Ni-catalyst. Reproduced with permission from ref. 28. © Korean Society of Industrial and Engineering Chemistry.

of the counter separator whilst the substrates were fed from the top. The reaction would occur within the separator and product would be bled from the bottom of the column. The ionic liquid containing the catalyst would then be recycled from the top to the bottom of the reactor. The conditions throughout the reactor would be the same and the product catalyst separation would be integrated with the reaction. To our knowledge, such a process has never been described.

Two genuinely continuous flow processes involving ionic liquids have been, however, suggested and demonstrated. One is based on supporting a thin film of ionic liquid within a porous support and to use it in a tubular heterogeneous reactor—so-called supported ionic liquid phase (SILP) catalysts.[29] The other is to transport substrates and products into and out of the ionic liquid by using a supercritical fluid as the transport vector.[30–32]

5.4.2.2 Supported Ionic Liquid Phase Catalysis

Continuous flow hydroformylation of long chain alkenes (1-octene) has been carried out in a supported homogeneous system using a SILP catalyst, where a thin layer of ionic liquid catalyst is immobilised at the surface of an inert solid support (*e.g.* silica gel).

SILP catalysts have the ability to combine advantages from both homogeneous and heterogeneous catalysts. By using a fixed bed reactor in a homogeneous system, it is possible to achieve a better catalyst/product separation and still expect the same performance in terms of selectivity as a 'normal' homogeneous catalyst.[29,33]

SILP catalysis can be carried out in the liquid[29,34–37] or gas[28,37–40] phases. When using liquid phase substrates, there is a possibility that the catalyst and ionic liquid may leach from the support, especially if the ionic liquid has been designed to dissolve the substrate so as to reduce problems with mass transport. One way of getting around this problem is to chemically anchor the ionic liquid to the support. Mehnert and co-workers[29] investigated the hydroformylation of 1-hexene using $Rh/[BMMIM]_3[P(3\text{-}C_4H_6SO_3)_3]$ (BMMIM = 1-butyl-2,3-dimethyl imidazolium) and a series of ionic liquids supported in the pores of silica from which the surface hydroxyl groups had been removed by reaction with an imidazolium modified triethoxysilane. This gives a layer of imidazolium salt anchored to the silica, which may aid in the immobilisation of the ionic liquid. In general, it was found that turnover frequencies ($< 55\text{--}65\,h^{-1}$) were low compared with the ones for an analogous homogeneous system using Rh/PPh_3 ($400\,h^{-1}$), but higher than the ones for the bulk biphasic system ($23\ h^{-1}$). Leaching of the ionic liquid and the catalyst was significant but less when very polar ionic liquids such as $[BMIM]PF_6$ were used or if the conversion of the non-polar alkenes to the more polar aldehydes was kept to a minimum. Despite the leaching, the SILP process provides a very elegant solution to the catalyst product separation problem.

Hydrogenation of alkenes does not produce a polar product, so a similar system to that described above for the hydroformylation of liquid alkenes was used in the hydrogenation of 1-hexene to give enhanced activity compared with the homogeneous or bulk biphasic systems without any measurable leaching. The catalyst could be reused 18 times without any loss of activity.[41] Continuous reactions in a flow reactor were not reported.

One other problem that can arise when using SILP catalysts for liquid phase substrates in reactions which also involve gases is severe gas depletion. The aspect ratio of the pores within the support is high, so gases have to diffuse a long way to arrive at the catalytically active centres. Once the gas that is initially dissolved in the substrate has been consumed, the reaction rate will drop significantly as it becomes gas transport limited.

None of these problems occur if the reaction is carried out in the gas phase. Here, diffusion to the catalytic sites is fast and, being essentially involatile, neither the catalyst nor the ionic liquid is transported out of the reactor. Thus, continuous flow propene[33] or butene hydroformylation in supported [BMIM][n-$C_8H_{17}OSO_3$][42] hydroformylations have been performed over a SILP catalyst for several hundred hours using, for example, a Rh/sulfo-xantphos catalyst. The support must first be treated to remove acid sites that protonate the ligand but, under optimised conditions, the system works very well. There is a small fall of in rate at longer reaction times because aldol condensation products of the C_4 aldehydes formed block the pores and/or catalyst, but these products can be removed by evacuating the catalyst at elevated temperature. Following this treatment, the catalyst returns to its initial activity and selectivity.[33,37,38,40,43–45]

The main problem with using all gas phase reactions is their limited scope. There are rather few reactions, methanol carbonylation being another that has been investigated[36] where the substrates and products are all in the gas phase at the reaction temperature. An alternative is to use high temperatures and very low flow rates of the less volatile substrates. However, the total throughput of such a system will be rather low.

One other alternative is to carry out the SILP catalysis using relatively low volatility substrates transported over the catalyst bed dissolved in a super-critical fluid.[46] This type of system has a number of potential advantages. Gas diffusion is fast in the supercritical phase. With the substrates and gaseous reagents all being contained in the one phase, they all have very good access to the pores of the support and hence the catalytic centres. By using ionic liquids and ionic catalysts that are insoluble in the supercritical fluid, leaching should be minimised. Supercritical carbon dioxide ($scCO_2$) has been shown to increase the solubility of permanent gases within ionic liquids,[47–49] so that the rate of transport of gases into the ionic liquid should be improved. Finally, any heavy products that form may be soluble in the supercritical fluid so that fouling of the catalyst can be avoided.

A system for the hydroformylation of 1-octene using a SILP catalyst consisting of a rhodium complex formed *in situ* from [Rh(acac)(CO)$_2$] (acacH = 2,4-pentanedione) and [PrMIM][Ph$_2$P(3-$C_6H_4SO_3$)] (PrMIM = 1-propyl-3-methylimidazolium) dissolved in a thin film of [OctMIM][Tf$_2$N] (OctMIM = 1-octyl-3-methyimidazolium, Tf = CF_3SO_2) with CO_2 flow has

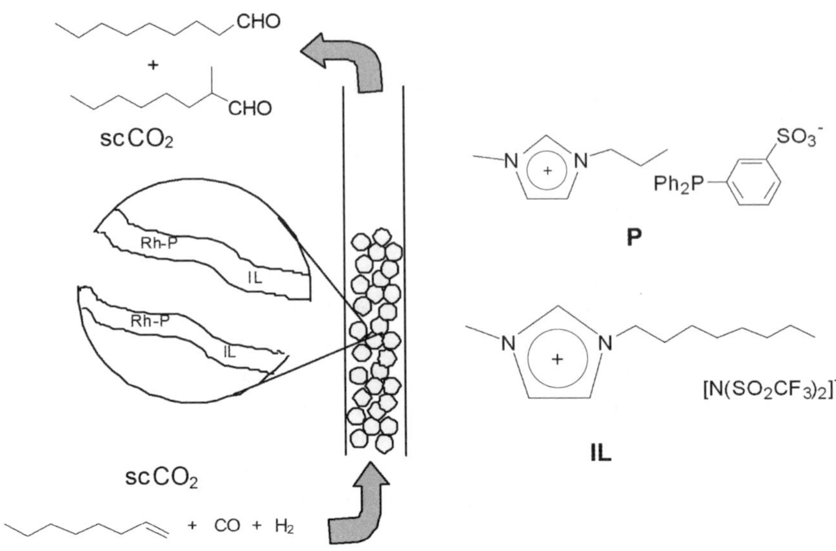

Figure 5.9 Schematic diagram of SILP catalyst with supercritical flow for the hydroformylation of 1-octene.[46]

been demonstrated and is shown schematically in Figure 5.9.[46] The reaction works best close to the critical point of the mobile phase producing over 500 catalyst turnover h^{-1} continuously for 40 hours, with Rh leaching of only 0.5 parts per million (ppm). Aldol condensation products from the C_9 aldehydes were also detected in the collected products, suggesting that fouling will not be a problem of long-term use.

5.4.2.3   *Ionic Liquids/scCO₂ Biphasic Systems*

The precursor to the SILP catalysts with supercritical flow were the $IL/scCO_2$ biphasic systems, which are amongst the most recent and promising alternatives to deal with the catalyst/product separation problem. These systems are based on the separation of the product and the retention of the catalyst in the solvent. The $scCO_2$ transports the products from the reactor, and they can then be precipitated from the solution after the decompression of the gas.[30–32]

Having been involved in showing that $scCO_2$ can be very soluble in some ionic liquids (up to 60 mol %), whilst the same ionic liquids have no measurable solubility in CO_2,[50] Blanchard and Brennecke described how it is possible to recover organic products successfully from the ionic liquid [BMIM][PF₆] using $scCO_2$.[51] They consider this separation technique as being quite reliable and environmentally friendly. Although this is a biphasic system, both phases are liquid. However, it is helpful that the $scCO_2$ is highly soluble in the ionic solution. As for the ionic solution, it has to be insoluble in the $scCO_2$ to permit product separation and catalyst retention.

Brennecke and co-workers studied the high-pressure phase behaviour of these systems.[52] They tested the solubility of carbon dioxide in ten different imidazolium salts, varying both the anion and the size of the alkyl chain. The results revealed that, for all ionic liquids tested, the solubility of the carbon dioxide increased with increasing pressure and decreased with increasing temperature. Its solubility also increased when using ionic liquids with fluoroalkyl groups present in the anion, such as $[Tf_2N]^-$, or with [methide]. As for the alkyl chain of the cation, the solubility of the carbon dioxide increased with increasing length of the chain.

Sellin *et al.* (hydroformylation of long chain alkenes)[31] and Leitner, Wasserscheid and co-workers (asymmetric hydrovinylation of styrene with ethene)[53] were the first to describe continuous flow in $IL/scCO_2$ biphasic systems, although others had shown that catalytic reaction products could be extracted from an ionic liquid reaction medium using $scCO_2$[54] and that reactions in ionic liquids could be carried out in the presence of $scCO_2$. The products could be removed and the ionic liquid containing the catalyst recycled.[55]

A brief explanation of the continuous flow hydroformylation process follows: the reactants (CO, H_2 and the alkene) and the CO_2 are introduced into the reactor in separate and continuous flows. Inside the CSTR, which already contains the ionic catalyst dissolved in the ionic liquid, the reactants become dissolved in the ionic liquid solution and react with the Rh catalyst. Both the ionic liquid and the catalyst are insoluble in the $scCO_2$, but the $scCO_2$ is soluble in the ionic liquid. The $scCO_2$ also increases the solubility of the CO and hydrogen in the ionic liquid.[47–49] The products dissolve in the $scCO_2$ and pass to the separator, where, by decreasing the pressure, the density of the CO_2 will also decrease and the products will precipitate from the solution into the collection vessel. The CO_2 can, in principle, be recycled (Figure 5.10)

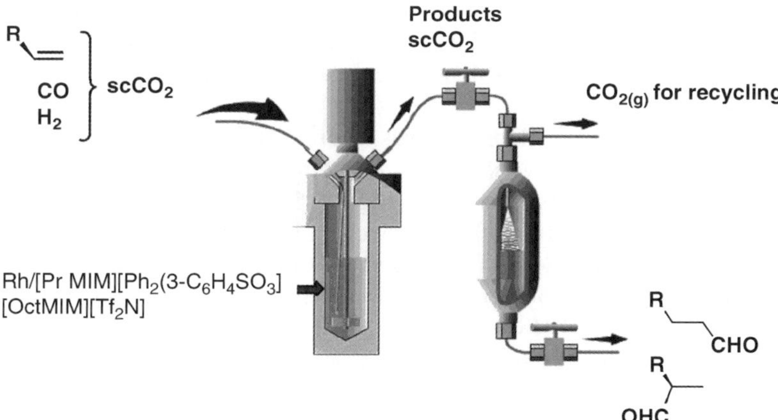

Figure 5.10 Schematic diagram of the continuous flow supercritical fluid–ionic liquid biphasic system applied in the hydroformylation of an alkene. After refs 32 and 56. © Royal Society of Chemistry, American Chemical Society.

The first experiments involved the use of Rh/[PrMIM][Ph$_2$P(3-C$_6$H$_4$SO$_3$)] [Ph$_2$P(3-C$_6$H$_4$SO$_3$Na)] was insufficiently soluble in the ionic liquid) in [BMIM]PF$_6$.[31] A stable reaction rate and linear selectivity were maintained over several hours, suggesting that the catalyst was stable, but the rate was very low (5 catalyst turnover h^{-1}). By increasing the length of the alkyl chain on the imidazolium salt and changing from [PF$_6$]$^-$ to [Tf$_2$N]$^-$ (desirable also because [PF$_6$]$^-$ reacts with water produced by aldol condensation of the product aldehydes to produce [O$_2$PF$_2$]$^-$ and HF),[31] the rate was increased by an order of magnitude[32] because the solubility of the 1-octene substrate in the ionic liquid was increased.[16] Further rate enhancements were achieved by varying the substrate flow, especially, the flow of CO/H$_2$.

Increasing the proportion of permanent gases in the flowing medium reduces the solubilising power of the CO$_2$ so that more of the substrate partitions into the ionic liquid, where the catalyst resides. In addition it makes the flowing medium a poorer solvent for the catalyst so that rhodium leaching, which was already low, is reduced still further. At relatively high levels of CO/H$_2$, the reaction becomes limited by the intrinsic catalyst kinetics rather than by mass transport effects, but care must be taken because the flowing medium may become such a poor solvent for the reaction products that it no longer extracts them properly, they build up in the reactor and eventually a liquid mixture containing substrate, products, catalyst and ionic liquid fills the separator, necessitating extensive cleaning and causing substantial downtime. Under the optimised conditions, steady state reaction rates of >500 h^{-1} (in the commercially interesting region) with very low rhodium leaching (12 ppb) were maintained over many hours.[32]

The main disadvantages of this process were the very high pressure required to extract the products from the reaction solution (200 bar) and the poor selectivity to the linear product ($\sim$3:1).[32] The first problem can be alleviated in the SILP system, described above, because there it is possible to work in the expanded liquid phase without forming a homogenous mixture and removing the catalyst and ionic liquid, or by removing the necessity for product extraction by dissolving the catalyst in the reaction product (see Section 5.4.1.6).[57,58]

The problem of product linearity was addressed by using a different ligand based on the highly selective xantphos core, modified to bear a pendant imidazolium functionality as shown in Figure 5.11, which gave linear (l) to branched (b) aldehyde ratios of 40 : 1 over an eight hour period with rates of 280 h^{-1}. Rhodium leaching was increased to about 0.2 ppm.[56]

The hydrovinylation of styrene was performed in a continuous flow system at 0 °C in [EMIM][Tf$_2$N] (EMIM = 1-ethyl-3-methylimidazolium) because the specially designed catalyst is thermally sensitive. Since the system was considerably below the critical temperature of CO$_2$, the reaction was probably carried out in liquid CO$_2$. Nevertheless, continuous operation was possible over >60 h with good conversion and 60–66 % enantiomeric excess (ee).[59]

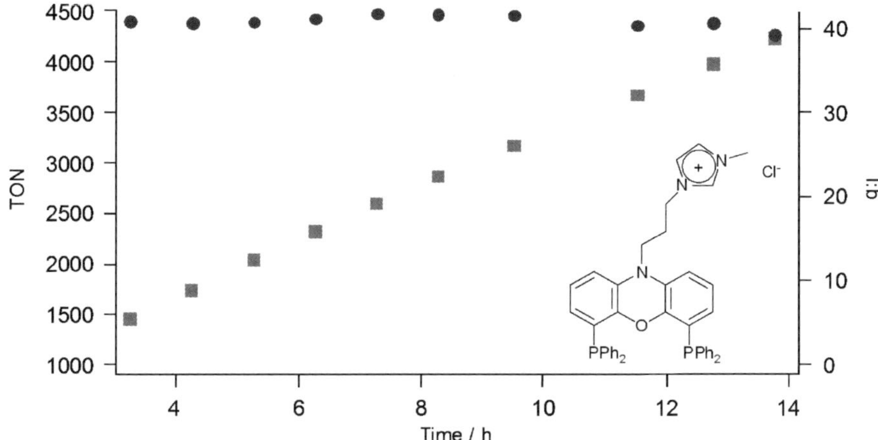

Figure 5.11 Hydroformylation of 1-octene in an scCO$_2$ [OctMIM][Tf$_2$N] system using a rhodium complex of the ligand shown. The selectivity to linear aldehyde was 92%. Reproduced with permission from ref. 56. © Royal Society of Chemistry.

This type of process has now been applied in a wide range of catalytic reactions (see the review in ref. 30).

5.5 Supercritical Fluids

A supercritical fluid (scF) is normally defined as a substance (highly compressed gas) above its critical temperature (T$_C$) and critical pressure (P$_C$), where the critical point represents the highest temperature and pressure at which the substance can exist as a vapour and liquid in equilibrium.[60]

Figure 5.12 shows the phase diagram for pure carbon dioxide which can, like many materials, exist in the solid, liquid or gaseous forms. The tie lines shown represent phase transitions and, in particular, that between the liquid and the gas represents evaporation or condensation. For non-ideal materials, the critical temperature, T$_c$, is the temperature above which the gas will not condense, no matter how high the pressure is raised. Above this temperature and above the critical pressure (the boiling pressure at T$_c$), the material is termed a supercritical fluid and it has some properties that resemble those of a liquid (densities >0.4 g cm^{-3}, ability to dissolve a range of organic compounds) and some that are gas-like (fills all the space available, flows like a gas, high diffusivity, total miscibility with permanent gases). In addition, the solubilising power of a supercritical fluid can be altered by changing the temperature and/or pressure. The critical point of CO$_2$ is 73.8 bar and 31.1 °C, conditions which are readily accessible in both research and industrial environments. This, along with the low toxicity, flammability and cost of CO$_2$ makes it the supercritical

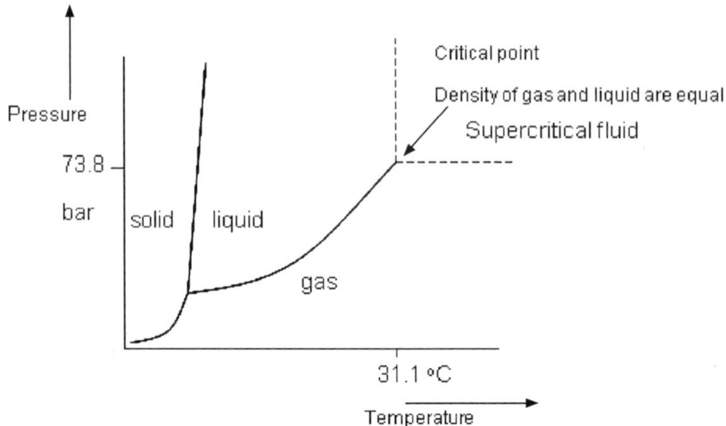

Figure 5.12 Phase diagram for pure carbon dioxide showing its critical parameters, $p_c = 73.8$ bar and $T_c = 31.1\,°C$.

Table 5.2 Critical parameters for several supercritical fluids. Reproduced with permission from ref. 60. © Wiley VCH.

Compound	T_c (K)	T_c (bar)
Carbon dioxide	304.1	73.8
Ethane	305.4	48.8
Ethene	282.4	50.4
Propane	369.8	42.5
Propene	364.9	46.0
Trifluoromethane (Fluoroform)	299.3	48.6
Chlorotrifluoromethane	302.0	38.7
Trichlorofluoromethane	471.2	44.1
Ammonia	405.5	113.5
Water	647.3	221.2
Cyclohexane	553.5	40.7
n-Pentane	469.7	33.7
Toluene	591.8	41.0

fluid of choice for many applications. The ability to dissolve organic materials at the same time as permanent gases as well as the excellent flow properties makes it an ideal medium for transporting compounds (even of quite low volatility) in flow systems.

Several other supercritical fluids have, however, also been studied for use in chemical processes. Table 5.2 shows the critical parameters for various compounds.[60]

There are many advantages associated with the use of supercritical fluids as reaction media,[61,62] but the main one is related to the environmental advantage of replacing common solvents as reaction media. Supercritical fluids also provide the added advantage of being able to release the products from the reaction medium by reducing the pressure in the system.

5.5.1 Supercritical Fluids for Product Separation in Homogeneous Catalysis

5.5.1.1 All Homogeneous Supercritical Systems

Two main approaches have been developed for using supercritical fluids to affect the separation of products from catalysts in supercritical systems. In one the catalyst is designed so that it is soluble in the $scCO_2$, whilst in the other the substrate and product are transported in the $scCO_2$ but the catalyst is insoluble.

In general, the substrates are soluble in $scCO_2$, so easier separation and adaptation to flow conditions occurs when the catalyst is insoluble. Nevertheless, many $scCO_2$ catalysts have been developed, either using alkyl phosphine ligands such as PMe_3[63] or PEt_3[64] or using fluorinated ponytails,[65] anions or both.[66,67] These all homogeneous systems can still be used for catalyst product separation because the catalyst is usually less soluble than the substrate in $scCO_2$. Using pressure and temperature swings it is possible to precipitate the catalyst in the reactor, remove the substrate dissolved in the supercritical fluid, precipitate the substrate in a separate vessel and recompress the $scCO_2$ for passage back into the reactor. This CESS Catalysis and Extraction using Supercritical Solution process has been demonstrated for the asymmetric hydroformylation of styrene using a rhodium complex with a fluorinated BINAPHOS ligand in a semi-batch process.[68,69] Yields and enantioselectivities were high, but some rhodium (0.2–2 ppm) and ligand loss occurred (as indicated by a drop in enantiomeric excess after the fourth cycle, which could be restored by the addition of extra ligand). A design concept for batch continuous operation has been reported.[70]

5.5.1.2 Supercritical Fluid–Water Biphasic Systems

One very elegant way of circumventing the pressure swings required in the CESS process whilst still using a CO_2 soluble catalyst is to work with reactions that give water-soluble products. Water is only poorly soluble in $scCO_2$. Such processes resemble coffee decaffeination because the product can be removed in the water phase without the need for decompression of the $scCO_2$, and hence very significant savings are obtained in energy utilisation.

Leitner and co-workers[66,67] used water–CO_2 systems to facilitate the separation of the catalyst from the product in homogeneous catalytic systems. By careful choice of the catalyst and the product, it is possible to have the catalyst dissolved in the $scCO_2$, with the product being preferentially soluble in water. Such reactions can be carried out in continuous flow mode with the substrate, reacting gases and water being continuous flowed into the reactor and the aqueous solution of the product being continuously removed. As for caffeine recovery, large pressure swings are not required so the process costs are reduced, but the number of products that fulfil the required criterion of preferential solubility in water is small.

One such reaction that has been demonstrated is the hydroformylation of itaconic acid or methyl-2-acetamido acrylate using a cationic rhodium complex of a BINAPHOS ligand bearing fluorous ponytails with BARF $\{B[3,5-C_6H_3(CF_3)_2]_4\}$ anions to render the catalyst completely CO_2 soluble (Scheme 5.2).[66,67] Excellent conversion and selectivity (ee = 93.6% *(S)* and 98.4% *(R)* respectively) were obtained. Using batch operation and removing the water phase after each run before adding more substrate and gases, the catalyst survived five repeat runs apparently unscathed.

Hancu and Beckman applied similar methodology in the homogeneous catalytic synthesis of hydrogen peroxide from H_2 and O_2 in CO_2.[71] This is a one step reaction catalysed by a CO_2-soluble palladium complex and is an alternative to its most popular commercial process, the anthraquinone or AQ process (Figure 5.13).

Although hydrogen peroxide is widely known as a green oxidant, the AQ process has quite a few drawbacks including the large energy input, the generation of waste and the number of unit operations in the process. Preliminary results have demonstrated the synthesis of hydrogen peroxide using both CO_2-soluble Pd(II) and Pd(0) catalysts bearing fluorinated phosphines, although the latter revealed better catalytic results. The product can be easily recovered from the reaction medium without any large drops in pressure because it can be extracted into water, which is only poorly soluble in $scCO_2$.

An added benefit is that the inert (non-oxidisable) nature of CO_2 means that handling oxygen is much safer than in conventional systems, where organic solvents and other material quickly exceed their explosion limits and severe risks are encountered at anything but very low oxygen partial pressures.

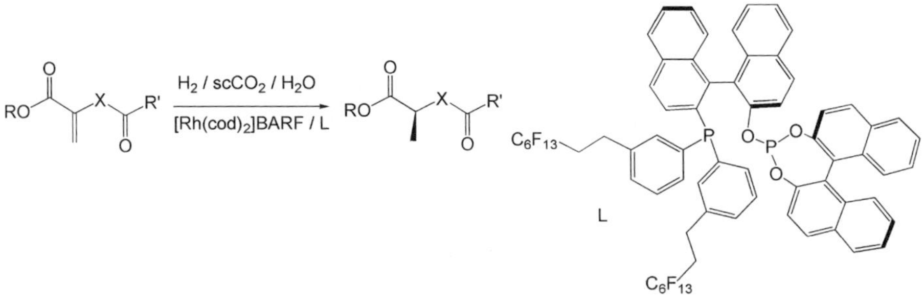

Scheme 5.2 Asymmetric hydrogenation of itaconic acid (R = H, R' = OH, X = CH$_2$) or of methyl-2-acetamido acrylate R = R' = Me, X = NH) a reverse phase CO$_2$/H$_2$O system catalysed by a cationic rhodium complex of fluorinated BINAPHOS. Modified and reproduced with permission from ref. 67. © Royal Society of Chemistry.

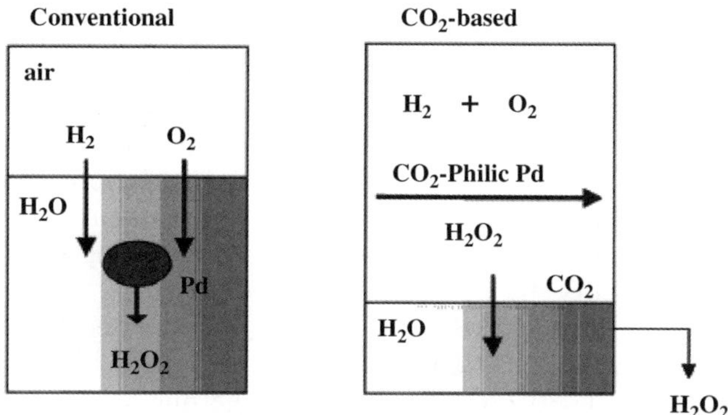

Figure 5.13 The AQ process *vs.* the direct reaction of hydrogen and oxygen for the production of hydrogen peroxide. Reproduced with permission from ref. 71. © Royal Society of Chemistry.

5.5.1.3 *Polymerisation in Supercritical Fluids*

One other area where scCO$_2$ soluble catalysts are important is when the product is insoluble in scCO$_2$ and precipitates during the reaction. A significant application of this concept is in polymerisation reactions, although the first examples made CO$_2$ soluble polymers. A partnership between DuPont and the University of North Carolina at Chapel Hill resulted in the first commercial example of fluoropolymer resin synthesis using carbon dioxide as a solvent in the Teflon^{TM} polymerisation process.[72,73] C$_2$F$_4$ is polymerised in scCO$_2$ using radicals derived from C$_4$F$_9$I as the initiator in a truly homogeneous process such that long-chain polymers can form. The use of CO$_2$ greatly increases the safety of the process[74] and the lifetime of the initiating radicals since, unlike almost all organic solvents, it is resistant to radical attack.[75]

Reactions where the polymeric product such as cross-linked poly(divinylbenzene) is insoluble in scCO$_2$ and precipitates during the reaction have been developed,[76,77] but continuous flow operation has not yet been demonstrated.

5.5.1.4 *Biphasic Systems Involving a Supercritical Fluid*

The majority of flow processes involving supercritical fluids as the transporting media that have been reported have the substrate, reacting gases and products dissolved in scCO$_2$ and the catalyst immobilised in some way so that it is insoluble. The immobilisation can be on a solid support or in a separate phase, the solvent for which has low solubility in scCO$_2$. In the biphasic systems, the catalyst is often rendered insoluble in scCO$_2$ by virtue of its being ionic, although sometimes high molecular mass is enough to immobilise the catalyst, for example, in a system where a rhodium hydroformylation catalyst is immobilised in excess 'polygard'—a mixture of tris(4-nonylphenyl)phosphites[78]—or in a hydrogenation reaction using [RhCl(PPh$_3$)$_3$] in polyethylene glycol.[79]

5.5.1.5 *Supported Catalysts with Supercritical Flow*

Early in 2002, Thomas Swan started operating a multipurpose flow plant using supercritical fluids (Figure 5.14). This plant has the potential to use $scCO_2$ as a solvent in various chemical processes such as hydrogenations, Friedel–Crafts alkylations, hydroformylations, etherifications and acylations. With the capacity to produce 1000 tonnes of products per year, this facility can easily work both commercially and as a pilot plant to continue testing the use of supercritical fluids as solvents.

The plant was constructed to develop the work initiated by Poliakoff and co-workers on catalytic reactions in supercritical fluids.[80–84] The selective hydrogenation of the C=C double bond in isopherone to give trimethylcyclohexanone (Scheme 5.3) was carried out with selectivities as high as those obtained in conventional solvents, but with higher rates, using a supported palladium catalyst and CO_2 as the transport medium.[83,84] Key parameters of the reaction are listed in Table 5.3 but of particular interest is the overall pressure of 40–60 bar, which greatly reduces the engineering and recycling costs

Figure 5.14 Thomas Swan plant for the use of supercritical fluids as solvents. Reproduced from www.sustain-ed.org/PAGES/Process/swan_co2.html © Thomas Swan and Co. Ltd.

Scheme 5.3 Hydrogenation of isopherone to 3,3,5-trimethylcyclohexane.[83,84]

Table 5.3 Comparison of optimised conditions for isopherone production in a supercritical plant in the laboratory and on the plant.[83,84] Reproduced with permission from ref. 83. © Royal Society of Chemistry.

	Laboratory scale	*Plant*
Reactor size	0.85 cm (i.d.), 25 cm	
Catalyst	2% Pd (supported)	2% Pd (supported)
T [°C]	56 (inlet) 100 (outlet)	104–116 (isothermal)
Pressure [bar]	40–60	40–60
Hydrogen [equivalents]	1.7–2.75	1.7
Substrate feed [wt %]	2–48	9–17
Rate [kg h^{-1}]	0.25	100
Selectivity to TMCH [%]	100	99.7

compared with those required when using $scCO_2$ (100–150 bar). The reaction is carried out in the expanded liquid phase, which has a number of important advantages in addition to the lower pressure of operation. Diffusion constants are higher, gases are more soluble and catalyst leaching is reduced compared with those for the liquid phase, whilst the substrate concentration is higher than for the gas phase reactions.

This process involved a heterogeneous catalyst, but Poliakoff and co-workers were also the first to introduce $scCO_2$ flow systems with supported homogeneous catalysts.[85] Using a ligand developed by van Leewen and co-workers[86–88] (see Figure 5.15), which had been shown to be highly robust for a variety of hydroformylation reactions in the liquid phase, Poliakoff and co-workers transported alkene, CO and hydrogen all dissolved in $scCO_2$ over the catalyst in a simple flow tube reactor. The overall pressure was 170 bar and the catalyst proved to be very robust in the continuous hydroformylation of 1-octene with very low rhodium leaching [<1.2 mg (mole product)$^{-1}$]. The activity (turnover frequency, TOF = 160 h^{-1})[85] was somewhat reduced compared with the liquid phase system (TOF = 287 h^{-1}),[86] but the selectivity to linear aldehyde remained high (94%, *cf.* 95% for the liquid phase system).

The activity is expected to be lower because the substrate, being dissolved in the mobile phase, is distributed throughout the reactor whereas, in the liquid phase, the substrate concentration at the catalytic sites is higher. This effect may be ameliorated somewhat by the better contact of the gases with the

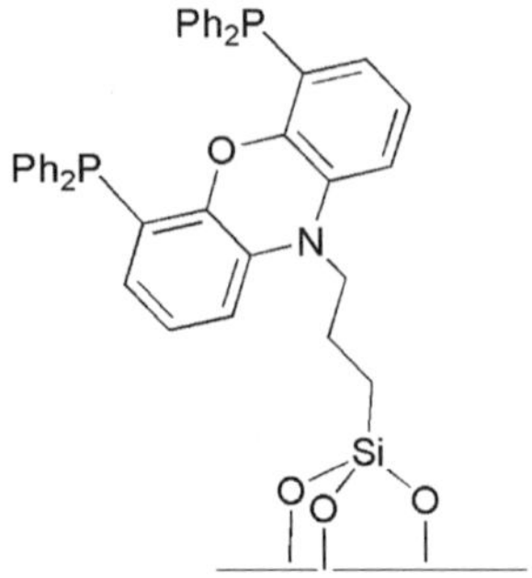

Figure 5.15 Ligand used for the hydroformylation of octane in a continuously flowing supercritical phase.[85,86] Reproduced with permission from ref. 2. © American Chemical Society.

Table 5.4 Comparison of methanol carbonylation using Rh/PVP with substrate in the liquid,[89] gas[90] or supercritical phases.[91a] Reproduced by courtesy of The Catalyst Group.

Phase	[MeI] mol dm^{-3}	T (°C)	P (bar)	TOF(h^{-1})	Rh leaching
Liquid ^{b}	1.07	150	40	280	40%^{c}
Gas ^{d}	1.37×10^{-3}	180	80^e (40) ^{f}	280	<1% ^{g}
scCO$_2$^{b}	0.107	150	200 (40) ^{f}	500	<0.08%

^{a}MeOH : MeI = 5, absolute concentrations vary enormously.
^{b}Batch reactions.
^{c}In 7.5 h.
^{d}Continuous flow.
^{e}1 : 1 CO : H$_2$.
^{f}CO pressure.
^{g}No losses detected by analysis of the solid.

catalyst in the CO$_2$ based system, but only if the transport of gases across the gas–liquid interface is rate-determining in the liquid phase system. The supported catalyst shown in Figure 5.15 is a very rare example where leaching in the liquid phase is undetectable and the lifetime is very long (one sample was used off and on for a year for different hydroformylation reactions without loss of activity or selectivity),[86] so it is difficult to improve on its performance.

However, using supercritical flow would be beneficial in cases where catalyst leaching into the liquid phase is significant, but reactivity is high. The lower solubilising power of the supercritical fluid should reduce the leaching to acceptable levels whilst retaining catalytic activity.

Results from one example are shown in Table 5.4 (though not in a flow system), which involves methyl iodide (MeI) promoted carbonylation of methanol using [Rh(CO)$_2$I$_2$]$^-$ supported on polyvinylpyrrolidone (PVP).

Kinetic studies of liquid phase reactions show that the catalyst leaching is very high (40% over a period of 7.5 h) and that most of the catalytic activity arises from dissolved (*i.e.* leached) catalyst, but not all.[89] Using $scCO_2$ as the solvent, the activity was almost doubled, but the catalyst could be reused six times without significant loss of activity. Rhodium in the recovered products was below the detection limit (0.3 ppm).[91] The same catalyst was used for all gas phase reactions, with the disadvantage that the space time yield was much reduced.[90]

In principle, any catalyst that can be immobilised on a solid support can be used for continuous flow catalytic reactions with supercritical transport, provided that the product is soluble in a supercritical fluid or can be transported as a CO_2 expanded liquid. We can thus expect to see major developments in this area, including commercialisation, over the next few years.

5.5.1.6 'Solventless' Systems with Supercritical Flow

In addition to the supercritical fluid–ionic liquid biphasic systems (Section 5.4.2.3) and other biphasic systems where the catalyst is dissolved in water (Section 5.5.1.2) or another liquid solvent (Section 5.5.1.4), it is possible to carry out homogenous reactions where no solvent is added but the substrates and products are carried through the reactor dissolved in $scCO_2$. Such continuous flow processes have been dubbed 'solventless' (Figure 5.16).[11]

Frisch *et al.* have described[11] such a system for the hydroformylation of 1-octene catalysed by a rhodium complex prepared *in situ* from [Rh(acac)(CO)$_2$] and [OctMIM][Ph$_2$P(3-C$_6$H$_4$SO$_3$)], which was chosen for its favourable solubility properties in the reaction product, nonanal, leading to high reaction rates.[58] The principle of these systems is quite similar to that used in the ionic liquid–supercritical fluid biphasic systems described previously in this chapter.

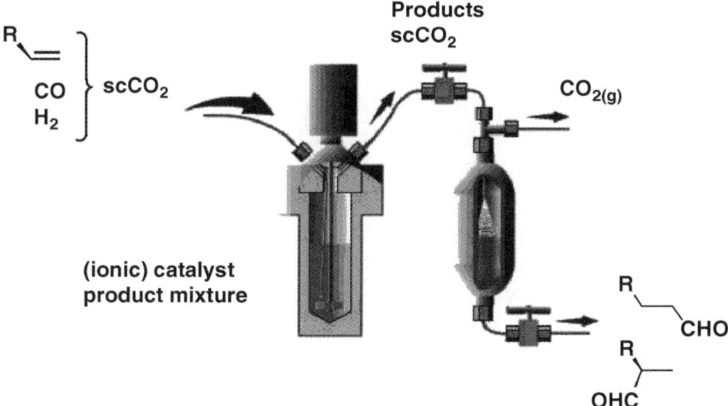

Figure 5.16 Schematic diagram for the continuous flow hydroformylation of alkenes, in which the catalyst is dissolved in the substrate/product mixture and the mobile phase is $scCO_2$. Reproduced with permission from ref. 11. © Royal Society of Chemistry.

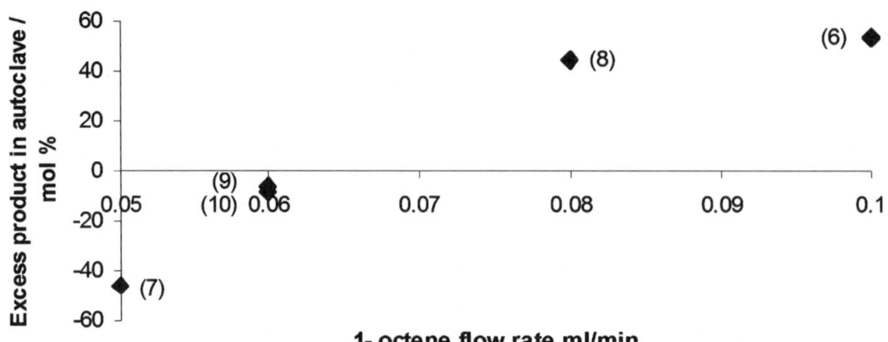

Figure 5.17   Plot showing the build-up or loss of material in the autoclave (mol of material recovered from the reactor – mol of material introduced at the beginning) as a function of 1-octene flow rate. CO_2 (2 nL min^{-1}), CO=H_2 (0.4 cm^3 min^{-1}), 100 °C, 140 bar. Normalised to a 10 h reaction time. Numbers in parenthesis are run numbers. Reproduced with permission from ref. 11. © Royal Society of Chemistry.

However, instead of using an ionic liquid as solvent, the catalyst is dissolved in the substrate–product mixture that develops during the reaction. The mobile phase used to extract the product from the reaction medium is $scCO_2$, which can, in principle, be recycled. These systems, besides being the most simple catalyst system possible—due to the absence of 'outsiders' except for the catalyst and the CO_2—have the potential to be run at lower pressure than the $scCO_2$–ionic liquid biphasic systems because the product does not require extraction from the ionic liquid, in which it is soluble.

The reactions are started by dissolving the catalyst precursors in a mixture of the starting material (1-octene) and product (nonanal) and the flows of substrate, gases (CO/H_2) and $scCO_2$ started. Care must be taken to balance the flow of substrate into the reactor, with the rate of extraction of the product from the reactor so that the level in the reactor remains constant. To help with this, a Jorgensen gauge can be incorporated into the reactor. Care must also be taken with the overall pressure because $scCO_2$ swells the liquid phase. If the ratio of CO_2 to CO/H_2 is too high, the liquid expands so much that it is forced out of the reactor taking the substrate, product **and catalyst** with it. Careful optimisation allowed suitable flow rates to be adopted (Figure 5.17) so that the reaction could be run continuously for at least 8 h with minimal rhodium leaching (0.3 ppm at the steady state), good activity (TOF = 180 h^{-1}), constant ratio of linear : branched aldehyde and a stable liquid level in the reactor (Figure 5.18).[11]

5.5.2   Recycling CO_2

One other issue that needs to be addressed in systems employing $scCO_2$ is CO_2 recycling. As already suggested, decompressing and recompressing CO_2 to allow

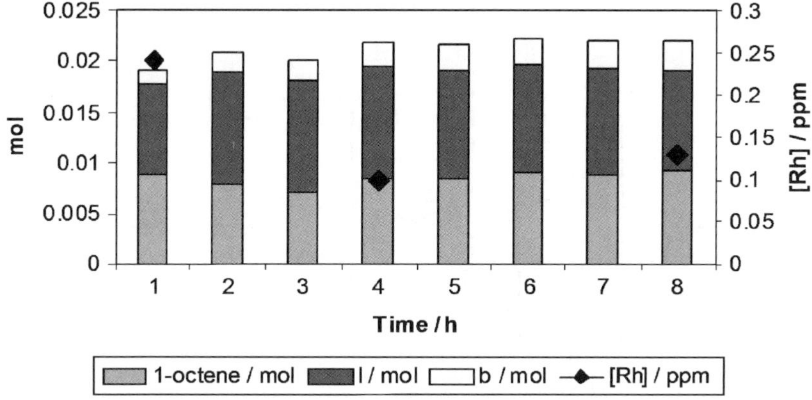

Figure 5.18 Composition of fractions collected under conditions of balanced flow from the 'solventless' hydroformylation of 1-octene. Conditions as for Run 10 in Figure 5.17. Reproduced with permission from ref. 11. © Royal Society of Chemistry.

for the separation of the products is an energy-intensive process. Extraction of the product at pressure with a solvent that is insoluble in $scCO_2$ is an option,[66,67] but then means that a further separation will be required. Often such solvents are not obvious. In principle, in the absence of a suitable extraction strategy, either recompression or cooling can be used before returning the CO_2 in the liquid phase to the reactor. Generally speaking, cooling below the critical temperature of 31.1 °C is the less energy-intensive option.

However, there is a further problem in reactions that employ permanent gases, especially as they are often used in excess over the amount of substrate. Either cooling or compressing a mixture of CO_2 containing permanent gases will lead to two phases. One is a liquid phase consisting predominantly of CO_2 but containing dissolved gases. This can be pumped back to the reactor, although outgassing on the way may constitute a problem. The second, more problematic phase is a gaseous phase consisting of CO_2 and permanent gases. If this is to be recycled, it must be compressed and if compression is carried out below the critical temperature, it will phase separate giving liquid and gaseous components which are difficult to pump. A design concept has been developed[56] in order to try to get around this problem (Figure 5.19). Here, the output from the continuous flow reactor passes through a heat exchanger (giving up some heat to the incoming gaseous fraction) to a separator where the product is collected by partial decompression. The gaseous phase (CO_2 and permanent gases) then passes to a chiller or compressor, where the majority of the CO_2 condenses. The gas phase then passes through the heat exchanger, where it is heated above its critical temperature and can then be compressed as a gas without condensation back into the reactor. The alternative, which completely circumvents the recompression issue, would be to vent the CO_2 coming out of the separator to atmosphere. It would be necessary to oxidise

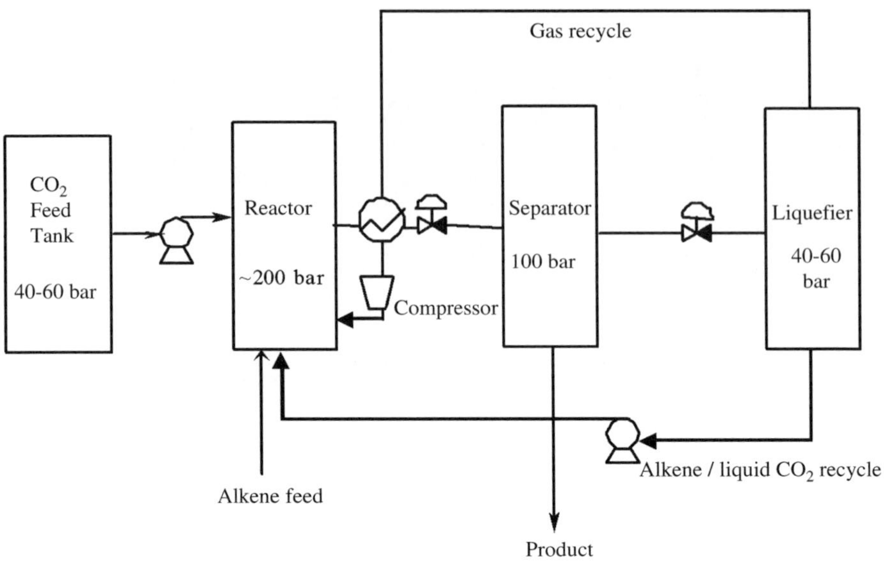

Figure 5.19 Process design for continuous operation with full recycling. Reproduced with permission from ref. 56. © Royal Society of Chemistry.

any unreacted permanent gases (CO, H_2) to CO_2 and water before venting. This may be a cheaper solution, but it is somewhat inelegant, does not hold up the CO_2 for long between production and loss to the atmosphere, wastes permanent gases and can generate small amounts of extra CO_2 (from CO oxidation).

5.6 Final Remarks

Concepts such as 'green chemistry', 'environmental chemistry', 'green solvents' or 'sustainability' are buzz words that have developed over the last few years. However, beneath their appealing exterior, there lurk very serious concerns about the development of modern ways of living. Every day, scientists from all research fields are being challenged by society to provide processes which make the products that make life so much safer and more comfortable, but in ways that will leave our planet with no adverse affects for future generations.

The chemical industry is charged with developing 'perfect chemical processes' in which the raw materials can be regenerated on the appropriate timescale and the process uses no energy, produces only the desired material, generates no waste and produces no pollution. Our goal must be to redesign existing chemical processes or design new processes to minimise the, until recently, normal formation of waste (often toxic and dangerous) and the use of hazard substances, which require special disposal after the process. In other words we need to develop processes with minimised E-factors,[92] with high atom economy and which fit with the twelve principles of green chemistry introduced by Anastas and Warner.[93] Catalysis provides an important step along the pathway to these ideals, and homogeneous catalysis with its high reaction selectivity and mild operating conditions is particularly attractive.

In this chapter, we have tried to outline processes for overcoming the main problems of homogeneous catalysis, that is:

- the separation of the reaction products from the catalyst and any solvent used in the reaction;
- the emission of volatile organic solvents as a result of solvent use;
- the requirement for batch or batch continuous processing.

In attempting to do this, we have highlighted flow systems which automatically build in the catalyst–product separation and often use supercritical carbon dioxide as the transport medium. Although CO_2 is a greenhouse gas (making the most significant contribution to global warming), it is not generated in the processes we describe, but is simply held up on its way from being generated to being released into the atmosphere. In all other ways (non-toxic, non-flammable, effective anti-explosion diluent), it is entirely benign in the environment. Its transport properties (high diffusivity), solubilising power (totally miscible with gases, good solvent for many organic compounds when pressurised) and low cost make it a highly suitable solvent for flow processes. The one disadvantage is the high pressures required to reach the critical condition. However, recent developments in the use of subcritical CO_2 and expanded liquids may go a long way towards addressing this potential problem. These types of systems work best if the catalyst is in the solid state either covalently attached to a support or dissolved in a thin film of a solvent, which is insoluble in $scCO_2$, supported on the inside of a porous solid.

The other type of solvent we have examined is ionic liquids. Having non-measurable vapour pressure, most ionic liquids do not pollute the atmosphere and in many cases they appear to have low toxicity. Full life-time analyses are lacking, but no major issues appear to be being thrown up. Ionic liquids can be designed to have all kinds of polarity and be miscible with different types of liquid, making them ideal for biphasic reactions, although continuous flow is best served by using them in tandem with supercritical fluids, either as biphasic systems or as the thin film for supported ionic liquid phase catalysts with or without supercritical flow.

The development of the use of these alternative solvents for catalytic processes is in its infancy. There is very much to be done, but they have the potential to create a major revolution not only in the chemical industry but also in society as a whole.

References

1. B. Cornils and W. A. Herrmann, *Applied Homogeneous Catalysis*, Wiley-VCH, Weinheim, 1996.

2. D. J. Cole-Hamilton, *Science*, 2003, **299**, 1702.
3. N. J. Ronde and D. Vogt, in *Catalyst Separation, Recovery and Recycling; Chemistry and Process Design*, ed. D. J. Cole-Hamilton and R. P. Tooze, Springer, Dordrecht, 2006, p. 73.
4. C. D. Frohning and C. W. Kohlpaintner, in *Applied Homogeneous Catalysis with Organometallic Compounds*, ed. B. Cornils and W. A. Herrmann, VCH, Weinheim, 1996, vol. 1, pp. 27.
5. P. W. N. M. Van Leeuwen and C. Claver, *Rhodium Catalysed Hydroformylation,* Kluwer, Dordrecht, 2000.
6. J. A. Banister and G. E. Harrison, *US Pat.*, 0186323, 2004.
7. D. J. Cole-Hamilton and R. P. Tooze, *Catalyst Separation, Recovery and Recycling: Chemistry and Process Design,* Springer, Dordrecht, 2006.
8. M. A. Abraham and L. Moens, *Clean Solvents: Alternative Media for Chemical Reactions and Processing,* American Society of Chemistry, Washington DC, 2002.
9. R. A. Sheldon, I. Arends and U. Hanefeld, *Green Chemistry and Catalysis,* Wiley-VCH, Weinheim, 2007.
10. FDA, http://www.fda.gov/cder/guidance/index.htm.
11. A. C. Frisch, P. B. Webb, G. Zhao, M. J. Muldoon, P. J. Pogorzelec and D. J. Cole-Hamilton, *Dalton Trans.*, 2007, 5531.
12. P. Wasserscheid and T. Welton, *Ionic Liquids in Synthesis,* Wiley-VCH, Weinheim, 2nd edn, 2008.
13. M. J. Earle and K. R. Seddon, *Workshop on Sustainable Chemistry,* Venice, Italy, 1998.
14. J. Dupont, R. F. de Souza and P. A. Z. Suarez, *Chem. Rev.*, 2002, **102**, 3667.
15. N. Jain, A. Kumar, S. Chauhan and S. M. S. Chauhan, *Tetrahedron*, 2005, **61**, 1015.
16. P. Wasserscheid and W. Keim, *Angew. Chem., Int. Ed.*, 2000, **39**, 3773.
17. T. Welton, *Coord. Chem. Rev.*, 2004, **248**, 2459.
18. P. J. Dyson and T. J. Geldbach, *Metal Catalysed Reactions in Ionic Liquids,* Springer, Dordrecht, 2005.
19. P. Wasserscheid and M. Haumann, in *Catalyst Separation, Recovery and Recycling; Chemistry and Process Design*, ed. D. J. Cole-Hamilton and R. P. Tooze, Springer, Dordrecht, 2005, p. 183.
20. G. Fitzwater, W. Geissler, R. Moulton, N. V. Plechkova, A. Robertson, K. R. Seddon, O. Jim Swindall and K. W. Joo, *Ionic Liquids: Sources of Innovation*, http://quill.qub.ac.uk/sources.
21. G. W. Parshall, *J. Am. Chem. Soc.*, 1972, **94**, 8716.
22. S. N. Falling and S. A. Godleski, *US Pat.*, 5 238 889, 1993.
23. S. N. Falling, J. R. Monnier, G. W. Phillips, J. S. Kanel and S. A. Godleski, in *21st Conference on Catalysis of Organic Reactions*, ed. S. R. Schmidt, Orlando, FL, 2006, pp. 327.
24. BASF, http://www.basf.com/group/corporate/en/innovations/innovation.
25. M. Maase, *Ionic Liquids in the Chemical Synthesis of Pharmaceuticals*, PharmTech.com, 1 October 2008 [online] http://pharmtech.findpharma.

com/pharmtech/Article/Ionic-Liquids-in-the-Chemical-Synthesis-of-Pharmae/ArticleStandard/Article/detail/552614, accessed 19 June 2009.

26. H. Olivier-Bourbigou and A. Forestiere, in *Ionic Liquids in Synthesis*, ed. P. Wasserscheid and T. Welton, Wiley-VCH, Weinheim, 2003, p. 258.

27. H. Olivier-Bourbigou and L. Magna, *J. Mol. Catal. A: Chem.*, 2002, **182**, 419.

28. P. Wasserscheid, *J. Ind. Eng. Chem.*, 2007, **13**, 325.

29. C. P. Mehnert, R. A. Cook, N. C. Dispenziere and M. Afeworki, *J. Am. Chem. Soc.*, 2002, **124**, 12932.

30. D. J. Cole-Hamilton, T. E. Kunene and P. B. Webb, in *Multiphase Homogeneous Catalysis*, ed. B. Cornils, Wiley VCH, Weinheim, 2005, Vol. **2**, p. 688.

31. M. F. Sellin, P. B. Webb and D. J. Cole-Hamilton, *Chem. Commun.*, 2001, 781.

32. P. B. Webb, M. F. Sellin, T. E. Kunene, S. Williamson, A. M. Z. Slawin and D. J. Cole-Hamilton, *J. Am. Chem. Soc.*, 2003, **125**, 15577.

33. A. Riisager, R. Fehrmann, M. Haumann and P. Wasserscheid, *Eur. J. Inorg. Chem.*, 2006, 695.

34. C. P. Mehnert, R. A. Cook, N. C. Dispenziere and E. J. Mozeleski, *Polyhedron*, 2004, **23**, 2679.

35. C. P. Mehnert, *Chem. Eur. J.*, 2005, **11**, 50.

36. A. Riisager, B. Jorgensen, P. Wasserscheid and R. Fehrmann, *Chem. Commun.*, 2006, 994.

37. A. Riisager, K. M. Eriksen, P. Wasserscheid and R. Fehrmann, *Catal. Lett.*, 2003, **90**, 149.

38. A. Riisager, R. Fehrmann, S. Flicker, R. van Hal, M. Haumann and P. Wasserscheid, *Angew. Chem., Int. Ed.*, 2005, **44**, 815.

39. A. Riisager, R. Fehrmann, P. Wasserscheid and R. van Hal, in *Ionic Liquids IIIb: Fundamentals, Progress, Challenges and Opportunities: Transformations and Processes*, ed. R. D. Rogers and K. R. Seddon, Oxford University Press, New York, 2005, **902**, p. 334.

40. A. Riisager, P. Wasserscheid, R. van Hal and R. Fehrmann, *J. Catal.*, 2003, **219**, 452.

41. C. P. Mehnert, E. J. Mozeleski and R. A. Cook, *Chem. Commun.*, 2002, 3010.

42. M. Haumann, K. Dentler, J. Joni, A. Riisager and P. Wasserscheid, *Adv. Synth. Catal.*, 2007, **349**, 425.

43. A. Riisager, R. Fehrmann, M. Haumann, B. S. K. Gorle and P. Wasserscheid, *Ind. Eng. Chem. Res.*, 2005, **44**, 9853.

44. A. Riisager, R. Fehrmann, M. Haumann and P. Wasserscheid, in *Proceedings of 12th International Symposium on Relations between Homogeneous and Heterogeneous Catalysis*, Florence, Italy, 2005.

45. A. Riisager, R. Fehrmann, P. Wasserscheid and R. van Hal, in Proceedings of *Symposium on Ionic Liquids: Fundamentals, Progress, Challenges and Opportunities* held at the 226th American-Chemical-Society National Meeting, New York, 2003.

46. U. Hintermair, G. Y. Zhao, C. C. Santini, M. J. Muldoon and D. J. Cole-Hamilton, *Chem. Commun.*, 2007, 1462.
47. M. Solinas, A. Pfaltz, P. G. Cozzi and W. Leitner, *J. Am. Chem. Soc.*, 2004, **126**, 16142.
48. A. Ahosseini, W. Ren and A. M. Scurto, *Ind. Eng. Chem. Res.*, 2009, **48**, 4254.
49. Z. Xie, W. K. Snavely, A. M. Scurto and B. Subramaniam, *J. Chem. Eng. Data*, 2009, **54**, 1633.
50. L. A. Blanchard, D. Hancu, E. J. Beckman and J. F. Brennecke, *Nature*, 1999, **399**, 28.
51. L. A. Blanchard and J. F. Brennecke, *Ind. Eng. Chem. Res.*, 2001, **40**, 287.
52. S. Aki, B. R. Mellein, E. M. Saurer and J. F. Brennecke, *J. Phys. Chem. B*, 2004, **108**, 20355.
53. A. Bösmann, G. Francio, E. Janssen, M. Solinas, W. Leitner and P. Wasserscheid, *Angew. Chem., Int. Ed.*, 2001, **40**, 2697.
54. R. A. Brown, P. Pollet, E. McKoon, C. A. Eckert, C. L. Liotta and P. G. Jessop, *J. Am. Chem. Soc.*, 2001, **123**, 1254.
55. F. C. Liu, M. B. Abrams, R. T. Baker and W. Tumas, *Chem. Commun.*, 2001, 433.
56. P. B. Webb, T. E. Kunene and D. J. Cole-Hamilton, *Green Chem.*, 2005, **7**, 373.
57. A. C. Frisch, P. B. Webb, G. Zhao, M. J. Muldoon, P. J. Pogorzelec and D. J. Cole-Hamilton, *Dalton Trans.*, 2007, 5531.
58. P. B. Webb and D. J. Cole-Hamilton, *Chem. Commun.*, 2004, 612.
59. A. Bosmann, G. Francio, E. Janssen, M. Solinas, W. Leitner and P. Wasserscheid, *Angew. Chem., Int. Ed.*, 2001, **40**, 2697.
60. P. G. Jessop and W. Leitner, *Reactions in Supercritical Carbon Dioxide*, Wiley-VCH, Weinheim, 1999.
61. P. G. Jessop and W. Leitner, *Chemical Synthesis using Supercritical Fluids*, Wiley-VCH, Weinheim, 1998.
62. E. J. Beckman, *J. Supercrit. Fluids*, 2004, **28**, 121.
63. P. G. Jessop, T. Ikariya and R. Noyori, *Nature*, 1994, **368**, 231.
64. M. F. Sellin, I. Bach, J. M. Webster, F. Montilla, V. Rosa, T. Aviles, M. Poliakoff and D. J. Cole-Hamilton, *J. Chem. Soc. Dalton Trans.*, 2002, 4569.
65. D. Koch and W. Leitner, *J. Am. Chem. Soc.*, 1998, **120**, 13398.
66. K. Burgemeister, G. Francio, V. H. Gego, L. Greiner, H. Hugl and W. Leitner, *Chem. Eur. J.*, 2007, **13**, 2798.
67. K. Burgemeister, G. Francio, H. Hugl and W. Leitner, *Chem. Commun.*, 2005, 6026.
68. G. Francio and W. Leitner, *Chem. Commun.*, 1999, 1663.
69. G. Francio, K. Wittmann and W. Leitner, *J. Organomet. Chem.*, 2001, **621**, 130.
70. C. M. Gordon and W. Leitner, in *Catalyst Separation, Recovery and Recycling: Chemistry and Process Design*, ed. D. J. Cole-Hamilton and R. P. Tooze, Springer, London, 2006, p. 215.

71. D. Hancu and E. J. Beckman, *Green Chem.*, 2001, **3**, 80.
72. T. J. Romack, J. R. Combes and J. M. De Simone, *Abstr. Pap. Am. Chem. Soc.*, 1995, **209**, 127.
73. T. J. Romack, J. M. De Simone and T. A. Treat, *Macromolecules*, 1995, **28**, 8429.
74. D. J. van Brunner, M. B. Shiflett and A. Yokozeki, *US Pat.*, 5 345 013, 1994.
75. J. M. De Simone, Z. Guan and C. S. Elsbernd, *Science*, 1992, **257**, 945.
76. A. I. Cooper, W. P. Hems and A. B. Holmes, *Macromol. Rapid Commun.*, 1998, **19**, 353.
77. W. P. Hems, T. M. Yong, J. L. M. van Nunen, A. I. Cooper, A. B. Holmes and D. A. Griffin, *J. Mater. Chem.*, 1999, **9**, 1403.
78. M. F. Sellin and D. J. Cole-Hamilton, *J. Chem. Soc. Dalton Trans.*, 2000, **11**, 1681.
79. D. J. Heldebrant and P. G. Jessop, *J. Am. Chem. Soc.*, 2003, **125**, 5600.
80. M. G. Hitzler and M. Poliakoff, *Chem. Commun.*, 1997, 1667.
81. M. G. Hitzler, F. R. Smail, S. K. Ross and M. Poliakoff, *Org. Process Res. Dev.*, 1998, **2**, 137.
82. M. G. Hitzler, F. R. Smail, S. K. Ross and M. Poliakoff, *Chem. Commun.*, 1998, 359.
83. P. Licence, J. Ke, M. Sokolova, S. K. Ross and M. Poliakoff, *Green Chem.*, 2003, **5**, 99.
84. P. Licence and M. Poliakoff, in *Multiphase Homogeneous Catalysis*, ed. B. Cornils, Wiley-VCH, Weinheim, 2005, Section 6.5.
85. N. J. Meehan, A. J. Sandee, J. N. H. Reek, P. C. J. Kamer, P. W. N. M. van Leeuwen and M. Poliakoff, *Chem. Commun.*, 2000, 1497.
86. A. J. Sandee, J. N. H. Reek, P. C. J. Kamer and P. W. N. M. van Leeuwen, *J. Am. Chem. Soc.*, 2001, **123**, 8468.
87. A. J. Sandee, R. S. Ubale, M. Makkee, J. N. H. Reek, P. C. J. Kamer, J. A. Moulijn and P. W. N. M. van Leeuwen, *Adv. Synth. Catal.*, 2001, **343**, 201.
88. A. J. Sandee, L. A. van der Veen, J. N. H. Reek, P. C. J. Kamer, M. Lutz, A. L. Spek and P. W. N. M. van Leeuwen, *Angew. Chem., Int. Ed.*, 1999, **38**, 3231.
89. N. De Blasio, M. R. Wright, E. Tempesti, C. Mazzocchia and D. J. Cole-Hamilton, *J. Organomet. Chem.*, 1998, **551**, 229.
90. N. De Blasio, E. Tempesti, A. Kaddouri, C. Mazzocchia and D. J. Cole-Hamilton, *J. Catal.*, 1998, **176**, 253.
91. R. J. Sowden, M. F. Sellin, N. De Blasio and D. J. Cole-Hamilton, *Chem. Commun.*, 1999, 2511.
92. R. A. Sheldon, *CHEMTECH*, 1994, **24**, 38.
93. P. T. Anastas and J. C. Warner, *Green Chemistry:Theory and Practice*, Oxford University Press, New York, 2000.

Subject Index

Figures are indicated in *italic* type, tables are indicated in **bold** type.